Sets, Sequences and Mappings

The Basic Concepts of Analysis

Kenneth W. Anderson
Dick Wick Hall

DOVER PUBLICATIONS, INC.
Mineola, New York

Bibliographical Note

This Dover edition, first published in 2009, is an unabridged republication of the work originally published in 1963 by John Wiley & Sons, Inc., New York.

Library of Congress Cataloging-in-Publication Data

Anderson, Kenneth W.
Sets, sequences and mappings : the basic concepts of analysis / Kenneth W. Anderson and Dick Wick Hall. — Dover ed.
p. cm.
Originally published: New York : Wiley, 1963.
ISBN-13: 978-0-486-47421-2
ISBN-10: 0-486-47421-6
1. Set theory. 2. Functor theory. I. Hall, Dick Wick, 1912– II. Title.

QA248.A66 2009
511.3'22—dc22

2009019371

Manufactured in the United States by Courier Corporation
47421601
www.doverpublications.com

Preface

A great deal has been said in recent years about the widening gap between calculus and advanced calculus. This gap might more appropriately be said to exist between "mechanical and intuitional" and "rigorous" mathematics. Regardless of terminology, the fact remains that many students who continue in mathematics beyond the calculus sequence find it a shocking experience, and are frustrated by the sudden shift in emphasis from the mechanical to the theoretical, from the concrete to the abstract.

We have written this book in the hope that it may help to bridge this gap. Although we rely on intuitional motivation wherever possible, we have tried to maintain a high degree of precision in the statements of definitions, axioms, and theorems, as well as rigor in the proofs. We have also attempted to handle carefully some of those elusive concepts which in calculus are declared "beyond the scope of this book," but which in advanced calculus and other higher level courses are often introduced casually by a statement such as "We assume that the reader is familiar with . . ." Many examples are used to clarify and illustrate new concepts as they are introduced. More than 300 problems, many of which amplify or supplement material in the text, are included. We have deliberately avoided sketches, simply because they sometimes can be as misleading as they are helpful. (How many students, for example, visualize a set of points on the real line as an interval merely because sets are so often illustrated in this fashion?) However, it is recommended that the student and the instructor use sketches freely in their work, bearing in mind the obvious limitations of such sketches.

The first five chapters consist of a systematic development of many of the important properties of the real number system plus careful treatment of such concepts as mappings, sequences, limits, and continuity. The introduction of open sets in Chapter III permits a simultaneous treatment of these concepts from a topological point of view without, however, mentioning the word "topology" in the first five chapters. Thus an analyst might consider this material as a course in "baby real variables," whereas a topologist might consider it as a development of the topology of the real line. The sixth and final chapter discusses metric spaces, and generalizes many of the earlier concepts and results to arbitrary metric spaces.

For review purposes, an index of axiom and key theorems is provided at the end of the book.

For the last three semesters, we have used this text, in its various stages of development, as the basis of a semester course at Harpur College, and have found it to be highly successful. This course is now a prerequisite for our two-semester advanced calculus sequence for mathematics majors, as well as for topology. Our experience has shown that the first five chapters can be covered in a three-hour course, leaving Chapter VI available for additional independent (or honors) work. The entire text could be covered comfortably in a four-hour course. Although intended primarily for sophomores who have completed the calculus sequence, this material has been given to some advanced undergraduates and beginning graduate students, who have found it both stimulating and challenging.

We would like to acknowledge the contributions, either direct or indirect, of a number of our colleagues, notably Professors Howard Alexander, Robert G. Bartle, Guilford Spencer, and Allen D. Ziebur, all of whom suggested improvements in our original manuscript. The concept of uniformly isolated sequences, used extensively in the text, was developed by Professor Norman Levine in a series of articles appearing in "The American Mathematical Monthly." We have drawn heavily on the material in *Elementary Topology*, by Hall and Spencer, published by John Wiley & Sons, and we wish to acknowledge Professor Spencer's co-authorship of this earlier textbook, as well as his excellent review of the present manuscript. Special thanks are due Professor Bartle for his careful review of our work, which resulted in the corrections of many errors and considerable improvement in the entire book.

The expert may wonder at our apparent preoccupation with sequences, particularly in proofs involving constructions and the Axiom of Choice. The treatment used herein is the culmination of many discussions between the authors and Professor Ziebur. At the latter's suggestion, we have introduced an Axiom of Choice for Sequences, and have made careful use of this axiom in many of our proofs. Some of these proofs seem a little complicated, especially for sophomores, and are starred for possible omission at the instructor's discretion. We are indebted to Professor Ziebur for his many helpful suggestions as well as for several of the proofs which he either created or improved.

We also wish to express our appreciation to the Division of Science and Mathematics, Harpur College, State University of New York, and to the staff of John Wiley & Sons for their tremendous help in the preparation of the manuscript.

October 1962

K. W. Anderson
Dick Wick Hall

Contents

IV

Convergence, 73

V

Continuity and Uniform Continuity, 97

VI

Metric Spaces, 142

I

Introduction to sets and mappings

1. Sets

Mathematics requires precision in the expression of abstract concepts and in the application of logical processes. This precision is attained through the use of special terminology and symbols which eliminate the normal ambiguity in everyday language. A mastery of this terminology and symbolism is essential to the student who expects to continue in the field of mathematics. For this reason, we have attempted to explain each symbol carefully, and to define each new concept precisely, labeling each as a definition for ease of reference. Many terms in everyday usage have a special and precise meaning in mathematics, and the student is cautioned not to confuse the normal usage with the precise mathematical meaning.

The term *set* is used to designate a collection of objects of some kind. These objects are called the *elements*, or *members*, or *points* of the set.

We generally designate sets by capital letters A, B, C, etc., and members of a set by small letters a, b, c, etc. If a set A consists of the points a, b, c, d, we write $A = \{a, b, c, d\}$. If A consists of just the one point a, we write $A = \{a\}$, thus distinguishing between the *point* a and the *set* $\{a\}$ consisting of only the point a. The notation "$a \in A$" means "a is a member of A," or "a belongs to A"; the notation "$a \notin A$" means "a does not belong to A."

Definition 1.1. The *universe* U is the totality of all points under consideration (during any investigation), and is the source from which we extract sets.

It is evident that the universe is itself a set, but it might be considered as "the master set"; that is, we restrict our horizon to the universe at hand, and do not recognize the existence of any other objects (or sets of objects) except those belonging to our universe. Thus, for example, the equation $x^2 + 1 = 0$ has no solution in the universe consisting of all real numbers. When we studied algebra, we found it necessary to enlarge our universe, and this led to a new universe consisting of all complex numbers.

Before stating our next definition, let us digress for a moment to discuss the term "if and only if," which in the sequel will be abbreviated "iff." Let α and β be two declarative statements. A typical theorem of mathematics is a statement of the form "If α is true, then β is true," which is often shortened to "If α, then β," or "α implies β." Mathematicians consider the following statements as equivalent; that is, any two of these statements have precisely the same meaning:

α implies β
α is a sufficient condition for β
β is a necessary condition for α
if α then β
β if α
α only if β

The typical definition in mathematics is a statement of the form "α is true iff β is true." Such a definition has the following equivalent forms:

α is true iff β is true
α is a necessary and sufficient condition for β
β is a necessary and sufficient condition for α
α implies β *and* β implies α

Mathematicians are pleased when they discover theorems which are "iff" statements, since any such theorem provides two equivalent descriptions of the same concept.

Definition 1.2. A set A is a *subset* of a set B iff every point of A is a point of B.

The statement of Definition 1.2 means that both of the following statements are true:

(1) If the set A is a subset of the set B, then every point of A is a point of B.

(2) If every point of the set A is a point of the set B, then A is a subset of B.

The notation $A \subset B$ (or equivalently, $B \supset A$) means "A is a subset of B," or "the set A is contained in the set B," or "B contains A."

It is clear from Definition 1.2 that for every set A, we must have $A \subset A$.

We have seen that the statement $A \subset B$ means that if $p \in A$, then $p \in B$. Thus, if A is not a subset of B (which we write $A \not\subset B$), then there must exist some point $p \in A$ such that $p \notin B$. Now, for any sets A and B, it is clear that one or the other of these conditions must be satisfied. We express this as follows.

Theorem 1.3. If A and B are sets, then either $A \subset B$ or $A \not\subset B$.

Consider the set $\{1, 2, 3, 4, 5\}$, and let it be our universe; that is, $U = \{1, 2, 3, 4, 5\}$. The subsets of the universe U are determined by taking a Gallup poll. For example, suppose we have a subset $K \subset U$, and we want to determine the members of K. We poll the members of U with the following results:

Is 1 in K?	Yes
Is 2 in K?	No
Is 3 in K?	No
Is 4 in K?	Yes
Is 5 in K?	Yes

Since we have polled our entire universe, we conclude that $K = \{1, 4, 5\}$.

Suppose we have another subset $H \subset U$, and here our Gallup poll yields a "no" from every member of U. We must then conclude that the set H contains no elements. For reasons that will become clear later, we still choose to consider H as a set, which we call the empty set and define as follows.

Definition 1.4. The *empty set* is the set containing no elements, and is denoted by ϕ.

Given any set A, it cannot be true that $\phi \not\subset A$, since there are no points in ϕ. Therefore, by Theorem 1.3, we have the following lemma.

Lemma 1.5. If A is any set, $\phi \subset A$.

Returning to the preceding example, suppose we have a subset $J \subset U$, and here our Gallup poll yields a "yes" from every member of U. We must then conclude that the sets J and U consist of exactly the same points, and are thus indistinguishable; that is, we have merely given the same set two different names. This leads us to the notion of equality of sets.

Definition 1.6. Two sets A and B are *equal* (written $A = B$) iff $A \subset B$ and $B \subset A$.

A reasonable question to ask at this point is: How many subsets can be extracted from a universe consisting of n points, where n is any positive integer? The answer lies buried in the technique of our Gallup poll. We have a total of n members to poll, and each member can give us one of two answers, "yes" or "no." If the answer is "yes," the point is in the subset; if the answer is "no," the point is not in the subset. Thus, in choosing subsets, we have two choices for each member of the universe: We can either *take it* as a member of the subset, or *leave it* out. Since our choice at any particular member is independent of our choice at each of the other members, we find that we have $2 \times 2 \times 2 \times \ldots \times 2$ (n factors) choices, or 2^n possible subsets. In our earlier example, where $U = \{1, 2, 3, 4, 5\}$, we have $2^5 = 32$ subsets. We could arrive at this result in still another way by considering the various subsets as combinations of "yes" answers in our poll, and using the theory of combinations from algebra. Remembering that the number of combinations of n things taken r at a time is given by the formula $C(n, r) = n!/[r!(n - r)!]$, we see that the number of subsets consisting of just one point corresponds to the number of combinations of just one "yes" among the five answers, or $C(5, 1) = 5!/(1!4!) = 5$. Similarly, the number of subsets consisting of two points is $C(5, 2) = 5!/(2!3!) = 10$; three points, $C(5, 3) = 10$; four points, $C(5, 4) = 5$; five points, (recalling that $0! = 1$ by definition), $C(5, 5) = 1$, where this single subset with five points is, of course, the set U itself. Finally, the number of subsets containing no points is $C(5, 0) = 1$, where this single set is the empty set. The total number of subsets is then $5 + 10 + 10 + 5 + 1 + 1 = 32$, which agrees with our preceding result.

Definition 1.7. The *intersection* of two sets A and B (written $A \cap B$) is the set of all points which are in both A and B.

Definition 1.8. The *union* of two sets A and B (written $A \cup B$) is the set of all points which are in at least one of the sets A and B.

To illustrate these concepts, let us consider an example.

Example 1.9. Let the universe be $U = \{1, 2, 3, 4, 5, 6, 7\}$, and suppose we have the following subsets:

$$A = \{1, 2, 5\} \qquad C = \{3, 5, 7\}$$
$$B = \{2, 3, 4\} \qquad D = \{1, 2, 4, 6\}$$

Then

$$A \cap B = \{2\}; \quad A \cap D = \{1, 2\}; \quad B \cap D = \{2, 4\}; \quad C \cap D = \phi$$

Also

$$A \cup B = \{1, 2, 3, 4, 5\}; \quad B \cup C = \{2, 3, 4, 5, 7\}; \quad C \cup D = U$$

We shall make free use of parentheses, much the same as in algebra. For example, Definitions 1.7 and 1.8 extend to more than two sets under the convention:

$$A \cap B \cap C = (A \cap B) \cap C; \qquad A \cap B \cap C \cap D = (A \cap B \cap C) \cap D$$

$$A \cup B \cup C = (A \cup B) \cup C; \qquad A \cup B \cup C \cup D = (A \cup B \cup C) \cup D, \text{ etc.}$$

Also, parentheses are essential when unions and intersections are combined, as in $A \cup B \cap C$. Here we may choose either the set $(A \cup B) \cap C$ or the set $A \cup (B \cap C)$, and these two sets are generally not equal. To see this, consider the universe U and the sets A, B, and C defined in Example 1.9. A little calculation (verify this) shows that $(A \cup B) \cap C = \{3, 5\}$, whereas $A \cup (B \cap C) = \{1, 2, 3, 5\}$.

We have seen that a set may be empty, and thus, if we are considering an arbitrary set A, we must allow for the possibility that $A = \phi$. (When we wish to exclude this possibility, we speak of a *nonempty set* A.) This brings up the following natural question: What happens if the empty set ϕ is involved in a union or an intersection? The answer follows immediately from Definitions 1.7 and 1.8, but because of its importance we state it for easy reference as a lemma.

Lemma 1.10. If A is any set (including $A = \phi$), then $A \cap \phi = \phi$ and $A \cup \phi = A$.

One of the important techniques of set theory is that of proving set equalities. Suppose, for instance, that we want to prove the following:

$$A \cap (B \cup C) = (A \cap B) \cup (A \cap C)$$

We might first test it by trying it on some particular sets. Let us again use Example 1.9. We see that $A = \{1, 2, 5\}$, and $B \cup C = \{2, 3, 4, 5, 7\}$, so that $A \cap (B \cup C) = \{2, 5\}$. On the other hand, $A \cap B = \{2\}$, and $A \cap C = \{5\}$, so that

$$(A \cap B) \cup (A \cap C) = \{2, 5\}$$

We have thus *verified* the equality for our particular choice of sets. Note, however, that we have not proved the statement, any more than we can prove the trigonometric identity $\sin 2x = 2 \sin x \cos x$ merely by verifying it for the particular choice $x = 0$. However, we cannot be sure that the statement is wrong either. Therefore it seems reasonable to try to prove it. The general method for proving equality of sets is indicated by Definition 1.6; that is, we show that each set is contained in the other. For simplicity of notation, let us designate the left-hand set above by L, and the right-hand set by R, so that we wish to prove $L = R$.

We choose an arbitrary point $p \in L$, and show that $p \in R$. Since the choice of p is arbitrary, we conclude that every point of L is also a point of R; that is, $L \subset R$. Similarly, we choose an arbitrary point $p \in R$, show that $p \in L$, and conclude that $R \subset L$; hence $L = R$. We now state the set equality as a theorem, and give a formal proof, following the preceding method. In this first proof, we have numbered each step, and we have also supplied reasons for each conclusion in the first half of the proof. This procedure is employed merely to clarify the technique, and is not continued in subsequent proofs.

Theorem 1.11 (The Distributive Law for Intersections over Unions). If A, B, C are sets, then

$$A \cap (B \cup C) = (A \cap B) \cup (A \cap C)$$

Proof. We prove first: $A \cap (B \cup C) \subset (A \cap B) \cup (A \cap C)$.

(1) If $A \cap (B \cup C) = \phi$, the result is trivial. (by Lemma 1.5)

(2) If $A \cap (B \cup C) \neq \phi$, let $p \in A \cap (B \cup C)$.

(3) Then $p \in A$ and $p \in (B \cup C)$. (by Definition 1.7)

(4) But $p \in (B \cup C)$ means $p \in B$ or $p \in C$. (by Definition 1.8)

(5) ***Case 1.*** If $p \in B$, then $p \in (A \cap B)$. (by Definition 1.7 and (3))

Therefore $p \in (A \cap B) \cup (A \cap C)$ (by Definition 1.8)

Case 2. If $p \in C$, then $p \in (A \cap C)$. (by Definition 1.7 and (3))

Therefore $p \in (A \cap B) \cup (A \cap C)$. (by Definition 1.8)

(6) Consequently $A \cap (B \cup C) \subset (A \cap B) \cup (A \cap C)$. (by steps (2) to (5) and Definition 1.2)

To complete the proof of the theorem, we must show that

$$(A \cap B) \cup (A \cap C) \subset A \cap (B \cup C)$$

(1) If $(A \cap B) \cup (A \cap C) = \phi$, the result is trivial.

(2) If $(A \cap B) \cup (A \cap C) \neq \phi$, let $p \in (A \cap B) \cup (A \cap C)$.

(3) Then, $p \in (A \cap B)$ or $p \in (A \cap C)$.

(4) ***Case 1.*** If $p \in (A \cap B)$, then $p \in A$ and $p \in B$.
Now $p \in B$ implies $p \in (B \cup C)$.
Therefore $p \in A \cap (B \cup C)$.

Case 2. If $p \in (A \cap C)$, then $p \in A$ and $p \in C$.
Now, $p \in C$ implies $p \in (B \cup C)$.
Therefore $p \in A \cap (B \cup C)$.

(5) Hence $(A \cap B) \cup (A \cap C) \subset A \cap (B \cup C)$.

Consequently $A \cap (B \cup C) = (A \cap B) \cup (A \cap C)$. QED

The second half of Theorem 1.11 can also be established in the following neat way, using Problem 8. We see at once that

$$B \subset B \cup C \quad \text{and} \quad C \subset B \cup C$$

By Problem 8,

$$A \cap B \subset A \cap (B \cup C) \quad \text{and} \quad A \cap C \subset A \cap (B \cup C)$$

Therefore $(A \cap B) \cup (A \cap C) \subset A \cap (B \cup C)$. QED

Theorem 1.12. Given any set A in the universe U, there is exactly one set X such that both of the following conditions hold:

$$(a)\ A \cup X = U \qquad (b)\ A \cap X = \phi$$

Proof. Suppose that there exist two such sets X and Y satisfying these conditions. Then

$$A \cup X = U, \qquad A \cap X = \phi, \qquad A \cup Y = U, \qquad A \cap Y = \phi$$

Consequently

$$Y = Y \cap U = Y \cap (A \cup X) = (Y \cap A) \cup (Y \cap X) = Y \cap X$$

Therefore, by Problem 3, $Y \subset X$. By a similar argument, $X \subset Y$. Therefore $X = Y$, and we have proven that there cannot be two sets satisfying the given conditions. There is at least one set satisfying these conditions, the set X defined as all points of U not in A. This completes the proof of the theorem. QED

We define the complement of a set A in the universe U to be the unique set X defined in Theorem 1.12. The word "unique" as used in mathematics means "one and only one."

Definition 1.13. The *complement* of a set A in the universe U is the unique subset X of U satisfying the two conditions

$$A \cup X = U \qquad A \cap X = \phi$$

We denote the complement of A by $C(A)$, and observe that $C(A)$ consists of all elements of U which are not in A.

It follows from Definition 1.13 that if U is the universe, then $C(U) = \phi$ and $C(\phi) = U$.

In general, when we are considering arbitrary sets, we may tacitly assume that some universe U is given, and that all the sets we are considering are subsets of U. Similarly, we may discuss the complement of a set, again assuming that we have taken the complement with respect to our universe U.

Theorem 1.14. For any set A, $C[C(A)] = A$.

Proof. We see from Definition 1.13 that $C[C(A)]$ is the unique set X satisfying both of the equations

$$C(A) \cup X = U \qquad C(A) \cap X = \phi$$

Since $X = A$ satisfies both these equations, we have at once $A = C[C(A)]$ QED

Theorem 1.15. If A and B are sets, then $A \subset B$ iff $C(B) \subset C(A)$.

Proof. Suppose first that $A \subset B$. By Problem 8,

$$A \cap C(B) \subset B \cap C(B) = \phi$$

so that $A \cap C(B) = \phi$. Consequently

$$\begin{aligned} C(B) &= C(B) \cap [A \cup C(A)] \\ &= [C(B) \cap A] \cup [C(B) \cap C(A)] = C(B) \cap C(A) \end{aligned}$$

Therefore, by Problem 3, $C(B) \subset C(A)$.

For the converse, suppose that $C(B) \subset C(A)$. Then by the preceding paragraph, $C[C(A)] \subset C[C(B)]$, which, by Theorem 1.14, can be written in the form $A \subset B$. This completes the proof.

QED

The next theorem states some important relationships among complements, unions, and intersections.

Theorem 1.16 (DeMorgan's Laws). If A and B are sets, then

(*a*) $C(A \cup B) = C(A) \cap C(B)$; (*b*) $C(A \cap B) = C(A) \cup C(B)$

Proof of (*a*). We prove that $C(A \cup B) \subset C(A) \cap C(B)$. It is easily seen that $A \subset A \cup B$ and $B \subset A \cup B$. Therefore, by Theorem 1.15, $C(A \cup B) \subset C(A)$ and $C(A \cup B) \subset C(B)$. Consequently

$$C(A \cup B) \subset C(A) \cap C(B)$$

To complete the proof of (*a*), we must show that

$$C(A) \cap C(B) \subset C(A \cup B)$$

By Lemma 1.6, we lose no generality if we assume that the left-hand set is non-empty. Let $p \in C(A) \cap C(B)$. Then $p \in C(A)$ and $p \in C(B)$, so $p \notin A$ and $p \notin B$. Thus $p \notin (A \cup B)$, and hence it follows that $p \in C(A \cup B)$. Therefore

$$C(A) \cap C(B) \subset C(A \cup B)$$

and (*a*) is proved.

The proof of (*b*) is left as an exercise.

Let us now consider an application of set theory where the universe does not consist of numbers. Let a and b be two statements, each of which in a given situation is either true or false. Let S be the set of all logical situations in which a and b occur;

that is, S is the universe in this case. We may then define the following sets:

$$\begin{aligned} A &= \text{set of all situations in which } a \text{ is true.} \\ C(A) &= \text{set of all situations in which } a \text{ is false.} \\ B &= \text{set of all situations in which } b \text{ is true.} \\ C(B) &= \text{set of all situations in which } b \text{ is false.} \end{aligned}$$

Now suppose we make the following *direct statement*, which we may assert as a theorem: "If a is true, then b is true." This says that, of all the situations in S, those for which a is true must be included among those for which b is true; or, in terms of the sets just defined, $A \subset B$. We can now apply Theorem 1.14 to obtain $C(B) \subset C(A)$. Using the same definitions, this is the same as the assertion: "If b is false, then a is false." This latter statement is the *contrapositive* of the given direct statement, and we have just indicated by means of set theory that *any direct statement is logically equivalent to its contrapositive statement.*

We now prove a simple theorem from number theory to illustrate the contrapositive technique of proof.

Theorem 1.17. If the square of a positive integer is even, then the positive integer is even.

Proof. The contrapositive of the given statement is as follows: If a positive integer is odd (that is, not even), then its square is odd. To prove this, we note that any positive integer which is odd can be written in the form $2k - 1$, where k is a positive integer. The square of this integer is then

$$(2k - 1)^2 = 4k^2 - 4k + 1 = 2(2k^2 - 2k) + 1$$

which is clearly an odd integer. QED

PROBLEMS

1. If $U = \{a, b, c, d\}$, find all the subsets of U.

2. Let $U = \{1, 2, 3, 4, 5, 6, 7, 8, 9, 10\}$, and let A, B, and H be defined as follows: $A = \{6, 8, 9\}$, $B = \{1, 3, 7, 8, 9\}$, $H = \{2, 6, 8, 9\}$. Then

(*a*) Find $A \cup B$ and $A \cap H$.

(*b*) Find $(A \cap B) \cup H$.

(*c*) Find $C(B)$.

(*d*) Find the set K defined as follows: The point x is in K iff $x + 3$ is in A.

3. Prove: $A = A \cap B$ iff $A \subset B$.
4. Prove: $A = A \cup B$ iff $B \subset A$.
5. Given that $A \cup B \subset A \cap B$, prove that $A = B$.
6. Prove the distributive law for unions over intersections; that is, show that for any three sets A, B, and C, $A \cup (B \cap C) = (A \cup B) \cap (A \cup C)$.
7. Use Lemma 1.10 to verify directly that the set equality in Theorem 1.11 is true in each of the following special cases:
(*a*) $A = \phi$.
(*b*) $B = \phi$.
(*c*) $C = \phi$.
8. Prove that if $A \subset B$ and if H is any set, then $A \cap H \subset B \cap H$ and $A \cup H \subset B \cup H$.
9. Prove the commutative laws for unions and intersections; that is, show that if A and B are any two sets, then

$$A \cup B = B \cup A \quad \text{and} \quad A \cap B = B \cap A$$

10. Prove the associative laws for unions and intersections; that is, show that if A, B, and C are any three sets, then

$$A \cup (B \cup C) = (A \cup B) \cup C \quad \text{and} \quad A \cap (B \cap C) = (A \cap B) \cap C$$

11. Prove part (*b*) of Theorem 1.16 in two different ways: first, by using points, and second, by letting $A = C(E)$, $B = C(F)$, and using part (*a*) of Theorem 1.16.
12. Prove part (*a*) of Theorem 1.16 by showing that the set $X = C(A) \cap C(B)$ satisfies both of the equations (1) $(A \cup B) \cap X = \phi$, and (2) $(A \cup B) \cup X = U$, where U is the universe. Why does this constitute a proof?

2. *Mappings*

Definition 2.1. Let H, K be sets, and f a rule that associates a unique element of K with each element of H. Then we write "$f: H \to K$" and say that f is a *mapping* of H *into* K. If $x \in H$, the point in K associated with x is denoted by $f(x)$, and is called the *image* of x under f. The set H is called the *domain of definition* of f, or simply the *domain* of f, and the set of images in K is called the *range* of f.

The preceding definition states that no element of H can map under f onto more than one element of K. However, several (or even all) elements of H may map onto the same element of K. We now illustrate some mappings.

Example 2.2. Let $H = \{1,2,3,4,5,6,7\}$, and $K = \{1,3,5,7,9\}$. We can define a mapping $f: H \to K$ by specifying the image under f of each element of H as follows:

$$\begin{array}{ll} f(1) = 3 & f(5) = 3 \\ f(2) = 5 & f(6) = 9 \\ f(3) = 7 & f(7) = 5 \\ f(4) = 3 & \end{array}$$

In this example, the domain of f is H, and the range of f is the set $\{3, 5, 7, 9\}$, which is a subset of K but not the entire set K.

Example 2.3. Let H and K be defined as in Example 2.2, and define f as follows:

$$\begin{array}{ll} f(1) = 3 & f(5) = 3 \\ f(2) = 3 & f(6) = 3 \\ f(3) = 3 & f(7) = 3 \\ f(4) = 3 & \end{array}$$

Here the domain of f is H, and the range of f is the set $\{3\}$ consisting of just the single point 3. In this case, we call f a constant mapping, which we define as follows.

Definition 2.4. If $f:H \to K$, and if for each $x \in H$, $f(x) = k$, where k is a given element of K, then f is called a *constant mapping*.

Example 2.5. Let H and K be defined as in Example 2.2, and define f as follows:

$$\begin{array}{ll} f(1) = 3 & f(5) = 3 \\ f(2) = 5 & f(6) = 9 \\ f(3) = 7 & f(7) = 1 \\ f(4) = 3 & \end{array}$$

Here, the domain of f is the set H, and the range of f is the entire set K. This illustrates the concept of an "onto" mapping, which we now define.

Definition 2.6. If $f:H \to K$, and if the range of f consists of the entire set K, then we say that f is a *mapping* of H *onto* K, and write "$f: H \xrightarrow{\text{onto}} K$."

Note that every "onto" mapping is an "into" mapping, but the converse is not necessarily true. What is the advantage of an "onto" mapping? Clearly, if $f: H \xrightarrow{\text{onto}} K$, we know that every element of K must be the image under f of at least one element of H.

Example 2.7. Let $H = K = \{1, 2, 3, 4, 5\}$, and define f as follows:

$$f(1) = 1 \qquad f(4) = 4$$
$$f(2) = 2 \qquad f(5) = 5$$
$$f(3) = 3$$

Here the domain and the range of f are the same set, and the image under f of each element is just the element itself. This characterizes the identity mapping, which we now define.

Definition 2.8. If f is the mapping of a set H onto itself defined by $f(x) = x$ for each $x \in H$, then f is called the *identity mapping on* H.

We now give a special application of onto mappings.

Example 2.9. Let K be a set of five gumdrops, which are colored red, blue, yellow, green, and white. Let H be the set whose elements are these five colors. Our aim is to tag each gumdrop with its color; so we define a mapping f from the set H of colors onto the set K of gumdrops as follows:

$$f(\text{red}) = \text{red gumdrop} \qquad f(\text{yellow}) = \text{yellow gumdrop}$$
$$f(\text{white}) = \text{white gumdrop} \qquad f(\text{green}) = \text{green gumdrop}$$
$$f(\text{blue}) = \text{blue gumdrop}$$

Now let us abbreviate the colors red, blue, yellow, green, and white by r, b, y, g, and w, respectively, and let k stand for gumdrop. Then our colored gumdrops can be denoted by k_r, k_b, k_y, k_g, and k_w. Note that our mapping becomes

$$f(r) = k_r \qquad f(y) = k_y \qquad f(w) = k_w$$
$$f(g) = k_g \qquad f(b) = k_b$$

More generally, let $H = \{r, b, y, g, w\}$, and $K = \{k_w, k_y, k_b, k_g, k_r\}$. Note that we have defined a mapping $f: H \xrightarrow{\text{onto}} K$ by

$$f(x) = k_x \text{ for each } x \in H$$

We say that the set K has been indexed by the set H under the mapping f; f is called the index mapping, and H is called the index set.

Definition 2.10. If $f:H \xrightarrow{\text{onto}} K$, and if for each $\alpha \in H$ we define the image of α under f in K by $f(\alpha) = k_\alpha$, we say that the set $K = \{k_\alpha\}$ has been *indexed by the set H* under the mapping f. The mapping f is called the *index mapping*, and the set H is called the *index set.*

Our next aim is to introduce a very important index set, but we must first digress for a brief discussion of the positive integers. We first encountered them when we learned to count, and in this process we called them off in their order of increasing magnitude. As a result, when we think of the positive integers, we automatically imagine them as arrayed in this order: 1, 2, 3, 4, 5, 6, . . . , which we call their *natural ordering*. The positive integers form a set, but the point we want to emphasize is that they form a set regardless of any question of order. We could imagine, for example, that each positive integer is written on a separate card, and all the cards are thrown into a basket. Then the basket contains the set of all positive integers, which we denote by I. When we want to consider the set of positive integers with their natural ordering, we denote it by I_o. (The subscript o here might be thought of as meaning "ordered.") For easy reference, we define these notations.

Definition 2.11. We use the following notation:

$I = \{1, 2, 3, 4, \ldots\}$ is the set of all *positive integers.*

$I_o = \langle 1, 2, 3, 4, \ldots \rangle$ is the set of all *positive integers with their natural ordering.*

We adopt the convention that whenever we have a mapping $f: I_o \xrightarrow{\text{onto}} A$, the set A becomes an ordered set, inheriting its order from I_o. For any such mapping f, we define $a_n = f(n)$. The manner in which A inherits its order from I_o is illustrated in the following diagram:

$$\begin{array}{ccccccc} \langle 1, & 2, & 3, & \ldots, & n, & \ldots \rangle \\ \downarrow & \downarrow & \downarrow & & \downarrow & \\ \langle a_1, & a_2, & a_3, & \ldots, & a_n, & \ldots \rangle \end{array}$$

Note carefully that the first element of I_o maps onto the first element of A; the second element of I_o maps onto the second element of A; and, in general, the nth element of I_o maps onto the nth element of A. This is precisely what is meant by the statement, "A inherits its order from I_o."

Definition 2.12. A *sequence* is a set that has been indexed by the set I_o, and which inherits its order from I_o.

It is clear from this definition that every sequence is an ordered set, but it should be equally clear that every set is not a sequence. In order for a set to be a sequence, there must be an ordering of the points of the set inherited from the natural ordering of the positive integers under an index mapping.

Definition 2.13. Let A be a sequence and

$$f: I_o \xrightarrow{\text{onto}} A$$

the corresponding index mapping. The element $f(n)$ of A obtained as the image of the positive integer n under this mapping is called the *nth term* (or *general term*) of the sequence, and is denoted by a_n. Frequently, the sequence is denoted by $\langle a_n \rangle$.

Theorem 2.14. I_o is a sequence.

Proof. The result is immediate, since in this case, the index mapping is merely the identity mapping; that is, we define $f: I_o \xrightarrow{\text{onto}} I_o$ as follows: For each $n \in I_o$, $f(n) = n$. QED

We might note at this point that there are many ways (other than I_o) in which the set I of positive integers may be written as a sequence. The only requirement is the existence of an index mapping from I_o onto I. One such sequence is illustrated in the following example.

Example 2.15. Define $f: I_o \xrightarrow{\text{onto}} I$ as follows: For each $n \in I_o$,

$$f(n) = \begin{cases} n + 1 & \text{when } n \text{ is odd} \\ n - 1 & \text{when } n \text{ is even} \end{cases}$$

We could also write $f(n) = n - (-1)^n$ for each $n \in I_o$. Then, since 1 is odd,

$$f(1) = 1 + 1 = 2$$

since 2 is even, $f(2) = 2 - 1 = 1$

since 3 is odd, $f(3) = 3 + 1 = 4$, etc.

Thus, the mapping f determines the sequence $\langle 2, 1, 4, 3, 6, 5, \ldots \rangle$.

It is equally possible to define a sequence which is an ordered subset of I_o.

Example 2.16. Let A be the set of all positive even integers, and define $f: I_o \xrightarrow{\text{onto}} A$ as follows: For each $n \in I_o$, $f(n) = 2n$. Under this mapping f, A becomes the sequence $\langle a_n \rangle$, where for each n, $a_n = 2n$; or, we could write $A = \langle 2, 4, 6, 8, \ldots \rangle$.

If we compare the sequence A in Example 2.16 with I_o, we see that $A \subset I$, and that the members of A are arranged in the same order as their relative order in I_o. These two properties characterize a subsequence of I_o, which we now define.

Definition 2.17. The sequence $A = \langle n_i \rangle$ is a *subsequence of* I_o iff the following conditions hold:

(1) Each n_i is a positive integer (that is, $A \subset I$).

(2) For every positive integer i, $n_i < n_{i+1}$.

With the aid of Definition 2.17, we may now define the concept of a subsequence of any sequence. As might be expected, a subsequence of a sequence $\langle a_n \rangle$ is a sequence whose terms are chosen from the sequence $\langle a_n \rangle$, and arranged in the same order as their relative order in $\langle a_n \rangle$. Formally, we state this as follows.

Definition 2.18. The sequence $\langle b_i \rangle$ is a *subsequence of the sequence* $\langle a_i \rangle$ iff there exists a subsequence $\langle n_i \rangle$ of I_o such that $b_i = a_{n_i}$ for each positive integer i.

A few examples may help to clarify the notation.

Example 2.19. Let $\langle a_i \rangle$ be a sequence; that is, $\langle a_i \rangle = \langle a_1, a_2, a_3, a_4, \ldots \rangle$ and let $\langle n_i \rangle$ be a subsequence of I_o defined by $n_i = 3i$ for each i; that is, $\langle n_i \rangle = \langle 3, 6, 9, 12, \ldots \rangle$. We wish to determine the subsequence $\langle b_i \rangle$ of the sequence $\langle a_i \rangle$, where for each i, $b_i = a_{n_i}$. By the definition of $\langle n_i \rangle$, we see that $b_i = a_{3i}$ for each i. Thus, $b_1 = a_3$, $b_2 = a_6$, $b_3 = a_9$, etc. Therefore the subsequence $\langle b_i \rangle = \langle b_1, b_2, b_3, \ldots \rangle$ is $\langle a_3, a_6, a_9, \ldots \rangle$, which consists of every third term of the given sequence $\langle a_i \rangle$.

Example 2.20. Let $\langle a_i \rangle$ be the sequence $\langle 3, 6, 9, 12, 15, \ldots \rangle$; that is, $a_i = 3i$ for each positive integer i. (The sequence $\langle a_i \rangle$ is already a subsequence of I_o, but this fact does not concern us here.) Let $\langle n_i \rangle$ be the subsequence of I_o defined by $n_i = i^2 + 7$ for each i. Thus, the sequence $\langle n_i \rangle$ may be written as $\langle n_i \rangle = \langle 8, 11, 16, 23, 32, \ldots \rangle$. The corresponding sequence $\langle b_i \rangle$ defined by $b_i = a_{n_i}$ for each i must thus have $b_1 = a_8 = 24$,

$b_2 = a_{11} = 33$, $b_3 = a_{16} = 48$, $b_4 = a_{23} = 69$, and so on. Thus, $\langle b_i \rangle$ may be written as $\langle 24, 33, 48, 69, 96, \ldots \rangle$, which is surely a subsequence of $\langle a_i \rangle$.

A sequence may have the property that no two of its terms are the same. If so, we call it a sequence of distinct terms, which we define formally as follows.

Definition 2.21. If $\langle a_i \rangle$ is a sequence such that $a_i \neq a_j$ whenever $i \neq j$, then $\langle a_i \rangle$ is called a *sequence of distinct terms.*

All sequences defined explicitly in the previous examples are sequences of distinct terms. Lest this create the false impression that all sequences are of this type, we consider a few more examples. In the first one, we jump to the other extreme and define a sequence *all* of whose terms are the same.

Example 2.22. Let k be given, and choose as the index mapping the constant mapping; that is, for every $n \in I_o$, $f(n) = k$. We thus get the sequence $\langle k, k, k, k, \ldots \rangle$.

As a result of Example 2.22, we see that any set consisting of just one point may be written as a sequence, each of whose terms is the given point.

Example 2.23. Define the index mapping f as follows: For each $n \in I_o$,

$$f(n) = \begin{cases} 1 & \text{if } n \text{ is odd} \\ 0 & \text{if } n \text{ is even} \end{cases}$$

Note that f may also be written in the form $f(n) = [1 - (-1)^n]/2$ for each $n \in I_o$. This mapping yields the sequence

$$\langle 1, 0, 1, 0, 1, 0, 1, \ldots \rangle$$

Note that the sequence of Example 2.23 contains subsequences of the type in Example 2.22, with $k = 0$ and $k = 1$.

PROBLEMS

13. Let $H = \{1, 2, 3, 4\}$ and $K = \{1, 2, 3, 4, 5\}$. If $f: H \to K$ is defined by $f(x) = x + 1$ for each $x \in H$, what are the domain and range of f? Is f an onto mapping?

14. If $f: H \xrightarrow{\text{onto}} K$ is defined by $f(x) = x^3 + 1$, and if $K = \{2, 9, 28, 65\}$, determine H, where H is a subset of the real numbers. Why can we be sure of this?

15. Let A be the set of all positive integers which are divisible by 3. Is A a sequence? If not, can we write A as a sequence? If so, write A as a sequence in two different ways, and in each case indicate the index mapping f. (Be careful with the notation.)

16. Let A be the set of all positive multiples of 7. Then

(*a*) Write A as a sequence $\langle a_i \rangle$ in two different ways.

(*b*) If $\langle n_i \rangle$ is a subsequence of I_o defined by $n_i = i^2 + 7$ for each i, then for each sequence $\langle a_i \rangle$ in part (*a*), write the subsequence $\langle b_i \rangle$ determined by $\langle n_i \rangle$ according to Definition 2.18.

17. Let A be the set of all positive even integers, and B the set of all positive multiples of 4. Write A and B as sequences $\langle a_i \rangle$ and $\langle b_i \rangle$, respectively, with their natural ordering and then show that $\langle b_i \rangle$ is a subsequence of $\langle a_i \rangle$. Again, be careful with the notation.

18. If $\langle b_i \rangle$ is a subsequence of $\langle a_i \rangle$, and if $\langle c_i \rangle$ is a subsequence of $\langle b_i \rangle$, show that $\langle c_i \rangle$ is a subsequence of $\langle a_i \rangle$.

3. *Real numbers*

In both analytic geometry and calculus, we freely used the correspondence between the set of all real numbers and the set of points on a directed line (the X-axis), and used the terms "point" and "number" interchangeably. We also used the order relationships: Less than ($<$); equal to ($=$); greater than ($>$); less than or equal to ($\leq$); greater than or equal to ($\geq$). The real numbers arranged in their order of algebraically increasing values are referred to as the *real numbers with their natural ordering*.

Definition 3.1. We shall designate by R_1 the *set of all real numbers with their natural ordering* (or equivalently, the set of all points on the real line, directed positively from left to right).

From this point on, our universe is R_1, unless explicitly stated otherwise.

Definition 3.2. Let A be a set in R_1. Then b is an *upper bound* for A iff $x \leq b$ for every x in A.

Example 3.3. Let $A = \{1, 2, 3\}$. Then 3 is an upper bound for A, since for each $x \in A$, $x \leq 3$. Is 4 also an upper bound for A? Yes, since it is also true that for every $x \in A$, $x \leq 4$. In fact, any number greater than 3 is also an upper bound for A.

We thus see that upper bounds are not unique; that is, if A is any set and b is an upper bound for A, then for any $c > b$, c is also an upper bound for A. Note that an upper bound for a set may be a member of the set, but need not be. In Example 3.3, we saw that 3 is an upper bound for A, and $3 \in A$. However, 3 is the only upper bound for A which is a member of A. Let us consider one more example.

Example 3.4. Let A be the set of all real numbers x such that $0 < x < 1$. Then 1 (or any number larger than 1) is an upper bound for A. But $1 \notin A$. As a matter of fact, no upper bound for A is a member of A.

In the technique of contrapositive proof (see Theorem 1.17 and the discussion preceding it), we are interested in the negation of properties. Thus, for instance, we might ask "What do we mean when we say that a number b is not an upper bound for a set A?" The general process of negating definitions is subtle, and sometimes difficult, but the present case is relatively simple, as follows.

Theorem 3.5. The real number b is not an upper bound for the set A iff there exists an element x of A such that $x > b$.

It might be interesting to use this theorem and the contrapositive technique to *prove* the assertion in Example 3.4, that is, to prove that no upper bound for this particular set A can be a member of A. Thus, suppose that b is any point of A. By definition of the set A, this means that $0 < b < 1$. But since $b < (b + 1)/2 < 1$ (verify this), we see that $(b + 1)/2$ is a member of A which is greater than b. Hence, by Theorem 3.5, b is not an upper bound for A. Since b was originally chosen as any point in A, we have thus shown that if $b \in A$, then b is not an upper bound for A. This completes the proof.

Definition 3.6. Let A be a set in R_1. Then, b is a *lower bound* for A iff for every $x \in A$, $x \geq b$.

Just as for upper bounds, we see that lower bounds are not unique; that is, if b is a lower bound for A, then for any $c < b$, c is also a lower bound for A. Also, a lower bound for a set may, but need not, be a member of the set. Thus, in Example 3.3, 1 is a lower bound for A which belongs to A, whereas any number less than 1 is a lower bound for A which does not belong to A. In Example 3.4, the number zero (or any number less than zero) is a lower bound for A which does not belong to A.

Definition 3.7. A set is *bounded above* iff it has an upper bound. Similarly, a set is *bounded below* iff it has a lower bound. A set is *bounded* iff it is bounded both above and below. If a set is not bounded, we say that it is *unbounded.*

It follows from Definition 3.7 that an unbounded set fails either to be bounded above or bounded below, or both.

The empty set ϕ has the property that every number is an upper bound for ϕ, and every number is a lower bound for ϕ.

We now want to introduce one more useful set notation. For each x in our universe, let $P(x)$ be a statement, or assertion, or property which is either true or false. All those x for which $P(x)$ is true form a set. Thus, we may *define* a set in this way; that is, we may write the set as $\{x: P(x)\}$, which means "the set of all x such that $P(x)$ is true." For instance, the set A of Example 3.4 can now be written as $A = \{x: 0 < x < 1\}$.

Definition 3.8. For $x \in R_1$, the *absolute value of* x, written $|x|$, is defined as follows:

$$|x| = \begin{cases} x & \text{if } x \geq 0 \\ -x & \text{if } x < 0 \end{cases}$$

Thus, for example, since 2 is positive, $|2| = 2$; whereas since -2 is negative, $|-2| = -(-2) = 2$. The absolute value of a real number is therefore its numerical value, neglecting sign. If we interpret the real numbers as points on the real line, then absolute value has a geometrical significance; that is, $|x|$ merely denotes the *undirected distance* between the origin and the point x. Consider the following example.

Example 3.9. Let $A = \{x: |x| < 2\}$. Now, in terms of the geometrical significance of absolute value, we see that the set A consists of all those points whose undirected distance from the origin is less than 2. From this fact, we conclude that $A = \{x: -2 < x < 2\}$. Let us verify this result analytically, using the definition of absolute value. If $x \geq 0$, then $|x| = x$, so the condition on A for nonnegative x becomes $x < 2$. If $x < 0$, then $|x| = -x$, so the condition on A for negative x becomes $-x < 2$. Since multiplication of both sides of an inequality by a negative number reverses the sense of the inequality, this last inequality (after multiplying through by -1) becomes $x > -2$.

Combining the two results, we thus have $-2 < x < 2$ if $x \in A$; that is,

$$A = \{x: -2 < x < 2\}$$

Note that the same basic technique would yield a corresponding result if we were to replace 2 in Example 3.9 by any other positive number, and also if we were to allow for equality instead of the strict inequality. We thus have the following result, which we express as a lemma.

Lemma 3.10. For any $c > 0$,

$$\{x: |x| < c\} = \{x: -c < x < c\}$$

and for any $c \geq 0$,

$$\{x: |x| \leq c\} = \{x: -c \leq x \leq c\}$$

Just as $|x|$ denotes the undirected distance between the origin and the point x, so $|x - a|$ denotes the undirected distance between the point x and the point a.

Example 3.11. Let $A = \{x: |x - 3| < 2\}$. Here, we see that A consists of all those points whose undirected distance from the point 3 is less than 2. From this fact, we conclude that $A = \{x: 1 < x < 5\}$. We can again verify this analytically, but rather than use the definition of absolute value, we will use Lemma 3.10, replacing x by $x - 3$ and c by 2. This yields $-2 < x - 3 < 2$. Adding 3 to each term of these inequalities, we obtain

$$-2 + 3 < x - 3 + 3 < 2 + 3 \quad \text{or} \quad 1 < x < 5$$

the desired result.

Note that the same basic technique would apply in Example 3.11 if 3 were replaced by any real number, and 2 were replaced by any positive number, as well as if equality were allowed rather than strict inequality.

Lemma 3.12. Let $a, b \in R_1$.
Then, if $b > 0$,

$$\{x: |x - a| < b\} = \{x: a - b < x < a + b\}$$

and if $b \geq 0$,

$$\{x: |x - a| \leq b\} = \{x: a - b \leq x \leq a + b\}$$

Having arrived at this point, we may now look back and see that Lemma 3.10 is merely the special case of Lemma 3.12 where $a = 0$. It should also be evident that any set whose points satisfy an inequality of the form of those in Lemma 3.12 is necessarily bounded. In fact, if $A = \{x: |x - a| \leq b\}$, where $a, b \in R_1$ and $b \geq 0$, then Lemma 3.12 tells us that $a + b$ is an upper bound for A, and $a - b$ is a lower bound for A.

We now establish some of the important properties of absolute value, which we combine into a single theorem. Some of the proofs are carried out in complete detail to serve as a guide, others are sketched, and still others are left as exercises.

Theorem 3.13. Let $a,b \in R_1$. Then the following statements hold:

(1) $|-a| = |a|$.
(2) $-|a| \leq a \leq |a|$.
(3) $|ab| = |a| \cdot |b|$.
(4) $|a|^2 = a^2$
(5) $ab \leq |a| \cdot |b|$.
(6) $\left|\frac{a}{b}\right| = \frac{|a|}{|b|}$, provided $b \neq 0$.
(7) $|a + b| \leq |a| + |b|$ (Triangle Inequality).
(8) $|a - b| \leq |a| + |b|$.
(9) $|a - b| \geq \big||a| - |b|\big|$.

Proof. All nine statements follow easily if $a = 0$. We thus assume $a \neq 0$. Similarly, we may assume $b \neq 0$.

(1) If $a > 0$, then $-a < 0$; therefore $|a| = a$, and $|-a| = -(-a) = a$. If $a < 0$, then $-a > 0$; therefore $|a| = -a$, and $|-a| = -a$.

Thus, for every choice of a, $|-a| = |a|$.

(2) If $a > 0$, then $-|a| = -a < a = |a|$. If $a < 0$, then $-|a| = a < -a = |a|$.

(3) If $ab > 0$, then $|ab| = ab$. Also, a and b must have the same sign; so either both are positive, in which case $|a| = a$, $|b| = b$, and hence $|a|\,|b| = ab$; or both are negative, in which case $|a| = -a$, $|b| = -b$, and thus $|a|\,|b| = (-a)(-b) = ab$. If $ab < 0$, then $|ab| = -(ab)$. Also, a and b must have opposite signs. Without loss of generality, we may assume that $a > 0$ and $b < 0$, for otherwise

we could interchange the roles of a and b. Then $|a| = a$, and $|b| = -b$, so $|a|\,|b| = a(-b) = -(ab)$.

Thus for every choice of a and b, $|ab| = |a|\,|b|$.

(4) Follows immediately from (3) with $b = a$, since $|a^2| = a^2$.

(5) Follows immediately from (2) and (3).

(6) Left as an exercise.

(7) We shall use the contrapositive technique of proof here. Suppose that (7) is not true; that is, suppose that $|a| + |b| < |a + b|$ for some choice of a and b. Then since each side of this inequality is nonnegative, we can square each side to obtain

$$|a|^2 + 2|a|\,|b| + |b|^2 < |a + b|^2$$

which can be simplified by (4) to

$$a^2 + 2|a|\,|b| + b^2 < (a + b)^2$$

or

$$a^2 + 2|a|\,|b| + b^2 < a^2 + 2ab + b^2$$

which reduces to $|a|\,|b| < ab$. But this is the negation of (5). We have thus shown that if (7) is not true, then (5) is not true. But we know that (5) is true; hence (7) must be true.

(8) Write $a - b = a + (-b)$. Then replace b in (7) by $-b$ and use (1).

(9) Since $a = a - b + b$, $|a| = |a - b + b| \le |a - b| + |b|$ by (7). Hence

$$(i) \qquad |a| - |b| \le |a - b|$$

Also, since $b = b - a + a$, $|b| = |b - a + a| \le |b - a| + |a|$ by (7), so $|b| - |a| \le |b - a|$. But $|b - a| = |a - b|$ by (1). Thus we have $|b| - |a| \le |a - b|$, or, multiplying through by -1,

$$(ii) \qquad |a| - |b| \ge -|a - b|$$

Combining (i) and (ii) yields $-|a - b| \le |a| - |b| \le |a - b|$. By Lemma 3.10, with $x = |a| - |b|$ and $c = |a - b|$, we obtain

$$\big||a| - |b|\big| \le |a - b|$$

which is the desired result.

This completes the proof of Theorem 3.13.

With the use of absolute value, we can now establish a useful criterion for boundedness of a set.

Theorem 3.14. A set A is bounded iff there exists a nonnegative number M such that $|x| \leq M$ for every $x \in A$.

Proof. We show first that if $|x| \leq M$ for every $x \in A$, then A is bounded. By Lemma 3.10, $-M \leq x \leq M$ for every $x \in A$. Hence M is an upper bound for A, and $-M$ is a lower bound for A, so A is bounded. To complete the proof, we must show that if A is bounded, there exists M such that $|x| \leq M$ for every $x \in A$. Since A is bounded, A must be bounded both above and below. Let M_1 be an upper bound for A, and M_2 a lower bound for A; that is, $M_2 \leq x \leq M_1$ for every $x \in A$. Let M be the larger of the two numbers $|M_1|$ and $|M_2|$, which we normally indicate by writing $M = \max(|M_1|, |M_2|)$. Then, $-M \leq x \leq M$ for every $x \in A$, which by Lemma 3.10 says that $|x| \leq M$ for every $x \in A$. QED

PROBLEMS

19. State precisely, using proper notation, what is meant by: "b is not a lower bound for the set A."

20. Prove that no lower bound for the set A in Example 3.4 is a member of A.

21. For a given nonempty set A, define the set B by $B = \{x: -x \in A\}$. Prove that B is bounded below iff A is bounded above.

22. Show that a set cannot have two distinct upper bounds, both of which belong to the set.

23. Show that for $c \geq 0$, $\{x: |x| > c\} = \{x: x > c\} \cup \{x: x < -c\}$.

24. Express the set A in terms of inequalities involving x (with no absolute value symbols) if

(*a*) $A = \{x: |x - 2| \leq 1\}$.

(*b*) $A = \{x: 0 < |x - 2| \leq 1\}$.

25. Complete the proofs of (4), (5), and (8) of Theorem 3.13.

26. Prove (6) of Theorem 3.13.

27. Carry out an alternate proof of (7), Theorem 3.13, in the following way: Use (2), Theorem 3.13, on a and b separately, add the resulting inequalities, and then apply Lemma 3.10.

28. Show that if $|x - 2| < 1$, then $|(x - 2)(x + 5)| < 8$. Using the same method find a number $b > 0$ such that if $|x - 2| < b$, then $|(x - 2)(x + 5)| < 1$. (There are, of course, many such numbers. The idea here is to try to find one such number in a reasonable way, avoiding trial and error.)

29. Prove that a set A in R_1 is bounded iff A is empty or there exists a number $K \geq 0$ such that $|x - y| \leq K$ for every x, y in A.

4. *Suprema and infima*

We have discussed bounded sets of real numbers, and in particular have defined upper bounds and lower bounds for sets. We have seen that upper bounds and lower bounds are not unique; for example, if b is an upper bound for a set A, and if $c > b$, then c is also an upper bound for A. This fact leads quite naturally to the question: Among all the upper bounds for a set, is there a smallest one? If so, we call such a number the least upper bound for the set, or the supremum of the set (abbreviated lub and sup, respectively), and define it as follows.

Definition 4.1. If A is a set, then b is the *least upper bound* for A (written $b = \text{lub } A$), or the *supremum* of A (written $b = \sup A$), iff both the following conditions are satisfied:

(1) b is an upper bound for A.

(2) If $c < b$, then c is not an upper bound for A.

It is sometimes convenient to use condition (2) in its contrapositive form, which we list for easy reference.

(2′) If c is an upper bound for A, then $c \geq b$.

As with upper bounds in general, the supremum of a set may, but need not, be a member of the set. In Example 3.3, we let $A = \{1, 2, 3\}$. Here, $\sup A = 3$, and the supremum of the set is indeed a member of the set. In Example 3.4, we let $A = \{x: 0 < x < 1\}$. Then $\sup A = 1$, and the supremum of the set does not belong to the set.

Lemma 4.2. A nonempty set has at most one supremum.

Proof. Let A be a nonempty set, and suppose we have both $x = \sup A$ and $y = \sup A$. Then, by condition (1) of Definition 4.1, x and y are both upper bounds for A. But, by condition (2′) of Definition 4.1, $x = \sup A$ implies that $y \geq x$, and $y = \sup A$ implies that $x \geq y$. Therefore $x = y$, so there can be at most one supremum for A. QED

Now let us return to our original question: Among all the upper bounds for a set, is there a smallest one? In view of the terminology just developed, this question may be rephrased as follows: If a set has an upper bound, does it have a supremum, that is, a least upper bound? It is a fact, and can be proved in an axiomatic development of the real number system, that every nonempty set of real numbers which is bounded above has a least upper bound. But since a systematic treatment of the real numbers is a course in itself, we shall accept this very important property as an axiom.

Least Upper Bound Axiom. Every nonempty set of real numbers which has an upper bound has a least upper bound.

We could just as well have started this section with a discussion about lower bounds, and asked: Among all the lower bounds for a set, is there a largest one? Such a number, when it exists, is called the greatest lower bound for the set, or the infimum of the set, and is defined as follows.

Definition 4.3. If A is a set, then b is the *greatest lower bound* for A (written $b = \text{glb } A$), or the *infimum* of A (written $b = \inf A$), iff both the following conditions are satisfied:

(1) b is a lower bound for A.
(2) If $c > b$, then c is not a lower bound for A.

As with lower bounds in general, the infimum of a set may, but need not, be a member of the set.

In Example 3.3, where $A = \{1, 2, 3\}$, we have $\inf A = 1$, so $\inf A \in A$. In Example 3.4, where $A = \{x: 0 < x < 1\}$, we have $\inf A = 0$, and here, $\inf A \notin A$.

We leave as exercises for the student the statement of the contrapositive form of condition (2) and the proof that a nonempty set has at most one infimum. This still leaves us with the basic question: If a nonempty set of real numbers has a lower bound, does it have an infimum? The next theorem answers this question in the affirmative.

Theorem 4.4. Every nonempty set of real numbers that has a lower bound has a greatest lower bound.

Proof. Let H be a nonempty set of real numbers, and let b be a lower bound for H. Define the set K by: $K = \{p: -p \in H\}$. Let x be any element of K. Then, $-x \in H$, and since b is a lower bound for H, we must have $-x \geq b$. Thus, $x \leq -b$, and since x was chosen as any member of K, we see that $-b$ is an upper bound for K. By the Least Upper Bound Axiom, we may then conclude that the set K has a least upper bound. Let us denote it by $-c$, so that $-c = \sup K$. We now prove that c is the greatest lower bound for H; that is, $c = \inf H$. To do this, we show first that c is a lower bound for H. Let y be any member of H. Then, $-y \in K$. Since $-c = \sup K$, we must have $-y \leq -c$, which implies that $y \geq c$, and hence c is a lower bound for H. Now we must show that if $d > c$, then d is not a lower bound for H. Thus let $d > c$. Then $-d < -c$. But since $-c = \sup K$, we see that $-d$ is not an upper bound for K. Thus there exists $x_0 \in K$ such that $x_0 > -d$. This implies that $-x_0 < d$. However, $-x_0 \in H$ since $x_0 \in K$, and hence d is not a lower bound for H. Therefore, $c = \inf H$. QED

Combining the results of the Least Upper Bound Axiom and Theorem 4.4, we have the following important theorem.

Theorem 4.5. Every bounded nonempty set of real numbers has both a supremum and an infimum. Furthermore, if A is such a set, we always have $\inf A \leq \sup A$.

Theorem 4.6. Let H be a bounded nonempty set of real numbers, K the set of all upper bounds for H, and M the set of all lower bounds for H. Then

(i) K has an infimum, and $\inf K \in K$.

(ii) M has a supremum, and $\sup M \in M$.

Proof. By Theorem 2.5, H has both a supremum b and an infimum a.

We now prove (i). Clearly, b must be an upper bound for H, and hence $b \in K$. We want to show that $b = \inf K$. We first show that b is a lower bound for K. Thus, let x be any member of K. Then, by definition of the set K, x is an upper bound for H, and hence $x \geq b$. But this means that b is a lower bound for K. It remains to show that if $c > b$, then c is not a lower bound for K. Thus, let $c > b$. Then c is an upper bound for H, so $c \in K$. But we have already shown that $b \in K$, and since $c > b$, c cannot be a lower bound for K. Therefore $b = \inf K$. Note that we have proved not only that that $\inf K$ exists, but also that $\inf K$ is a member of K. The proof of (ii) is left as an exercise. QED

One of the most important applications of the Least Upper Bound Axiom is in the proof of the following theorem.

Theorem 4.7 (Archimedean Property of R_1). Given any $\epsilon > 0$, there exists a positive integer N such that $N\epsilon > 1$.

Proof. Suppose the theorem is not true. Then, for some $\epsilon > 0$, we have $n\epsilon \leq 1$ for every positive integer n; that is, the set $\{\epsilon, 2\epsilon, 3\epsilon, \ldots\}$ is bounded above. By the Least Upper Bound Axiom, this set must have a supremum, which we denote by c. Hence $n\epsilon \leq c$ for every $n \in I_o$. In particular, for each $n \in I_o$, we must have $(n + 1)\epsilon \leq c$, which reduces to $n\epsilon + \epsilon \leq c$, or $n\epsilon \leq c - \epsilon$. But this says that $c - \epsilon$ is an upper bound for the set. This is impossible, since $c - \epsilon < c$, and c is the supremum of the set. QED

The Archimedean property of R_1 says, in effect, that there is no such thing as an "arbitrarily small" positive number. It is true that our choice of a small positive number may be arbitrary; however, once it is chosen there is nothing arbitrary about its size. Theorem 4.7 tells us, in fact, that no matter how small the number may be, we can blow it up to any size we desire merely by taking a sufficient number of multiples. For example, given an $\epsilon > 0$, once we have found n from the theorem such that $n\epsilon > 1$, we could instead use $(2n)$, from which it follows that $(2n)\epsilon = 2(n\epsilon) > 2(1) = 2$. Similarly, if k is any positive number, $(kn)\epsilon > k$. In terms of geometry (where it was first used as an axiom), the Archimedean property asserts that any given scale, however small, can be used to mark off a distance which exceeds any given distance, however large. Thus, with a six-inch rule, or even a smaller one, we could measure the distance from New York to San Francisco.

Corollary 4.8. For any real number x, there exists a positive integer n such that $n > x$.

Proof. If $x \leq 0$, we may choose n as any positive integer, say $n = 1$. If $x > 0$, then $1/x > 0$, so by Theorem 4.7, there exists a positive integer n such that $n(1/x) > 1$. Thus, $n > x$. QED

Corollary 4.9. If a and b are positive numbers, there exists a positive integer n such that $na > b$.

Proof. By Theorem 4.7 (with $\epsilon = a$), there exists a positive integer N such that $Na > 1$. Let m_1 be an integer greater than N, so that $m_1 a > 1$. By Corollary 4.8 there exists a positive integer m_2 such that $m_2 > b$. Define $n = m_2 m_1$. Then, n is a positive integer (since it is the product of two positive integers). Furthermore, $na = (m_2 m_1)a = m_2(m_1 a) > m_2(1) = m_2 > b$. QED

Note carefully that if in Corollary 4.9 we let $n = N$, $a = \epsilon$, and $b = 1$, we then obtain Theorem 4.7 as a special case of this corollary; that is, Corollary 4.9 implies Theorem 4.7. Thus the three properties stated in the theorem and its corollaries are all equivalent.

PROBLEMS

30. State the contrapositive form of condition (2) in Definition 4.3.
31. Prove that a nonempty set has at most one infimum.
32. Prove (ii) of Theorem 4.6.
33. If the sets H, K, and M are defined as in the hypotheses of Theorem 4.6, under what condition could we have $\inf K = \sup M$?
34. Prove that if $A = \{x: 0 < x < 1\}$, then $\inf A = 0$ and $\sup A = 1$.
35. Given the set $\{\frac{2}{3}, \frac{4}{5}, \frac{6}{7}, \frac{8}{9}, \frac{10}{11}, \frac{12}{13}, \ldots, 2n/(2n+1), \ldots\}$.
(*a*) Find the least upper bound and greatest lower bound of this set.
(*b*) If b is the least upper bound which you find, then determine a member of the set which is greater than $b - \frac{1}{100}$.
36. Given the set K defined in Problem 35, prove that K may be made into the following sequence:

$$\langle \tfrac{2}{3}, \tfrac{4}{5}, \tfrac{6}{7}, \tfrac{8}{9}, \tfrac{10}{11}, \tfrac{12}{13}, \ldots, 2n/(2n+1), \ldots \rangle$$

You should do this by defining a mapping $f: I_o \xrightarrow{\text{onto}} K$. Basing your answer on the results of Problem 35, how would you define the infimum and supremum of this sequence?
37. Let A be a nonempty set, and define the set B by: $B = \{x: -x \in A\}$. Prove that $c = \sup A$ iff $-c = \inf B$.
38. Prove: If B is a bounded set of real numbers, and if A is a nonempty subset of B, then $\inf A \geq \inf B$ and $\sup A \leq \sup B$.
39. Give a direct proof that Corollary 4.8 implies Theorem 4.7.
40. Prove that if $b \in H$ and if b is an upper bound for H, then $b = \sup H$.
41. Prove that if $b \in H$ and if b is a lower bound for H, then $b = \inf H$.

II
Sequences

1. Monotone and bounded sequences

We saw in Chapter I that every sequence is a set, so our previous discussion on boundedness of sets applies automatically to sequences. However, to help clarify the general distinction between sets and sequences, we list a few of the basic concepts in terms of sequences. Throughout this chapter, unless otherwise stated, we continue to use the set R_1 of real numbers as our universe.

Definition 1.1. The real number b is an *upper bound* for the sequence $\langle x_n \rangle$ iff $x_n \leq b$ for every n. The real number b is a *lower bound* for the sequence $\langle x_n \rangle$ iff $x_n \geq b$ for every n. We say that the sequence $\langle x_n \rangle$ is: *bounded above* iff it has an upper bound; *bounded below* iff it has a lower bound; *bounded* iff it is bounded above and bounded below.

Theorem 1.2. The sequence $\langle x_n \rangle$ is bounded iff there exists a nonnegative number M such that $|x_n| \leq M$ for every n.

Proof. Same as proof of Theorem 3.14, Chapter I, with slight change in notation.

Theorem 1.3. Every subsequence of a bounded sequence is bounded; in fact, every upper bound for the sequence is an upper bound for every subsequence, and every lower bound for the sequence is a lower bound for every subsequence.

Proof. Left as an exercise.

Definition 1.4. A sequence $\langle x_n \rangle$ is *monotone nondecreasing* iff $x_n \leq x_{n+1}$ for every positive integer n. A sequence $\langle x_n \rangle$ is *monotone nonincreasing* iff $x_n \geq x_{n+1}$ for every positive integer n. A sequence is *monotone* iff it is monotone nondecreasing or monotone nonincreasing.

Note carefully what the preceding definition says. A monotone nondecreasing sequence is one in which successive terms may be the same or may increase, but can never decrease. For example, the sequence $\langle 1, 1, 2, 2, 3, 3, 4, 4, \ldots \rangle$ is monotone nondecreasing. In fact, even the sequence $\langle 1, 1, 1, 1, \ldots \rangle$ is monotone nondecreasing (it is also monotone nonincreasing) according to Definition 1.4. The sequence I_o is monotone nondecreasing, but in this particular case, we might be inclined to say that it is even better behaved than most such sequences, because equality never holds. It is precisely for this reason that we define a stronger flavor of monotonicity, eliminating any possible equality of terms.

Definition 1.5. A sequence $\langle x_n \rangle$ is *monotone increasing* iff $x_n < x_{n+1}$ for every positive integer n. A sequence $\langle x_n \rangle$ is *monotone decreasing* iff $x_n > x_{n+1}$ for every positive integer n. A sequence is *strictly monotone* iff it is monotone increasing or monotone decreasing.

Bear in mind that every monotone increasing sequence is also a monotone nondecreasing sequence. Thus any theorem which we prove for monotone nondecreasing sequences will also be true for monotone increasing sequences. Similarly, every monotone decreasing sequence is also a monotone nonincreasing sequence, and every strictly monotone sequence is a monotone sequence. Therefore, whenever possible, we shall simplify the statements of theorems by expressing them in their most general forms.

Now, as a result of Definition 1.5, we can say that the sequence I_o is monotone increasing. From the definition of a subsequence of I_o, we immediately have the following.

Theorem 1.6. Every subsequence of I_o is monotone increasing.

Theorem 1.7. (*a*) Every subsequence of a monotone nonincreasing sequence is monotone nonincreasing.

(*b*) Every subsequence of a monotone nondecreasing sequence is monotone nondecreasing.

(*c*) Every subsequence of a monotone increasing sequence is monotone increasing.

(*d*) Every subsequence of a monotone decreasing sequence is monotone decreasing.

Proof. We shall prove (*a*), leaving the proofs of (*b*), (*c*), and (*d*) as exercises. Let $\langle a_i \rangle$ be a monotone nonincreasing sequence, and let $\langle b_i \rangle$ be a subsequence of $\langle a_i \rangle$. Since $\langle a_i \rangle$ is monotone nonincreasing, we know that for any two positive integers i and j with $i < j$, $a_i \geq a_j$. In order to show that $\langle b_i \rangle$ is monotone nonincreasing, let k and m be two positive integers with $k < m$. We are given that $\langle b_i \rangle$ is a subsequence of $\langle a_i \rangle$, so there exists a subsequence $\langle n_i \rangle$ of I_o such that $b_i = a_{n_i}$ for each i. Hence, in particular, $b_k = a_{n_k}$ and $b_m = a_{n_m}$. But $k < m \in I_o$ implies that $n_k < n_m$, which, by the monotonicity of $\langle a_i \rangle$, says that $a_{n_k} \geq a_{n_m}$, or $b_k \geq b_m$. QED

It should be apparent that even though a sequence is not monotone, it may have a monotone subsequence. The sequence $\langle 1, 2, 1, 4, 1, 6, 1, 8, \ldots \rangle$ is not monotone, but it contains the monotone nondecreasing (or monotone nonincreasing) subsequence $\langle 1, 1, 1, 1, \ldots \rangle$, and it also contains the monotone increasing subsequence $\langle 2, 4, 6, 8, \ldots \rangle$. As a matter of fact, we now proceed to prove the startling fact that *every* sequence of real numbers *must* have a monotone subsequence.

First, however, we introduce two additional axioms for our work with the universe R_1. The first of these is an axiom for the positive integers which is equivalent to the principle of mathematical induction.

Well-Ordering Axiom for $\mathbf{I_o}$. The set I_o of positive integers is *well-ordered;* that is, every nonempty subset of I_o has a smallest element.

Our other axiom is a special case of what is known in mathematics as the Axiom of Choice.

Axiom of Choice for Sequences. Let $\langle A_n \rangle$ be a sequence of nonempty subsets of a universe U. Then there exists a sequence $\langle p_n \rangle$ of points of U such that given any $n \in I_o$, $p_n \in A_n$.

Given a sequence $\langle a_n \rangle$, we are interested in those particular subsequences of $\langle a_n \rangle$ that are determined by discarding the first N terms of the sequence $\langle a_n \rangle$, where N is any nonnegative integer.

The number N of terms discarded is indicated by a second subscript, as $\langle a_{n,N}\rangle$. If no term is discarded, we must have the original sequence, so that $\langle a_{n,0}\rangle = \langle a_n\rangle$. Thus a few such subsequences are as follows:

$$\langle a_{n,0}\rangle = \langle a_1, a_2, a_3, a_4, a_5, \ldots\rangle$$
$$\langle a_{n,1}\rangle = \langle a_2, a_3, a_4, a_5, \ldots\rangle$$
$$\langle a_{n,2}\rangle = \langle a_3, a_4, a_5, \ldots\rangle$$
$$\langle a_{n,3}\rangle = \langle a_4, a_5, \ldots\rangle; \qquad \text{and in general}$$
$$\langle a_{n,N}\rangle = \langle a_{N+1}, a_{N+2}, a_{N+3}, \ldots\rangle$$

Notice that each sequence in this list is a subsequence of all those sequences which precede it in the list; that is, for any nonnegative integer N, if k is a nonnegative integer, $\langle a_{n,N+k}\rangle$ is a subsequence of $\langle a_{n,N}\rangle$.

Lemma 1.8. Let $\langle a_n\rangle$ be a sequence such that for every nonnegative integer N, there exists an element of the sequence $\langle a_{n,N}\rangle$ which is a lower bound for the sequence $\langle a_{n,N}\rangle$. Then $\langle a_n\rangle$ has a monotone nondecreasing subsequence.

Proof. We are given that some term of the sequence $\langle a_n\rangle$ is a lower bound for $\langle a_n\rangle$. (This is the case $N = 0$.) Thus there exists (by the Well-Ordering Axiom) a smallest nonnegative integer k_1 such that a_{k_1} is a lower bound for $\langle a_n\rangle$. Define $b_1 = a_{k_1}$. Now let us consider the sequence $\langle a_{n,k_1}\rangle = \langle a_{k_1+1}, a_{k_1+2}, \ldots\rangle$. By hypothesis, some member of the sequence $\langle a_{n,k_1}\rangle$ is a lower bound for the sequence $\langle a_{n,k_1}\rangle$. Thus there exists a smallest nonnegative integer k_2 such that a_{k_2} is a lower bound for $\langle a_{n,k_1}\rangle$. Define $b_2 = a_{k_2}$. Since $\langle a_{n,k_1}\rangle$ is a subsequence of $\langle a_n\rangle$, we know that $b_1 \leq b_2$, by Theorem 1.3. Next we work with the sequence $\langle a_{n,k_1+k_2}\rangle$ to get a term a_{k_3} which is a lower bound for $\langle a_{n,k_1+k_2}\rangle$; defining $b_3 = a_{k_3}$, we have, by Theorem 1.3, $b_2 \leq b_3$. In general, suppose that for any positive integer j, we have determined the terms $b_1 \leq b_2 \leq \ldots \leq b_j$, where for each integer i such that $1 \leq i \leq j$, $b_i = a_{k_i}$. Then, by hypothesis, some member of the sequence $\langle a_{n,k_1+k_2+\ldots+k_j}\rangle$ is a lower bound for the sequence $\langle a_{n,k_1+k_2+\ldots+k_j}\rangle$. Thus there exists a smallest nonnegative integer k_{j+1} such that $a_{k_{j+1}}$ is a lower bound for the sequence $\langle a_{n,k_1+k_2+\ldots+k_j}\rangle$. Define $b_{j+1} = a_{k_{j+1}}$. By Theorem 1.3, $b_j \leq b_{j+1}$. Therefore we have defined inductively the entire sequence $\langle b_n\rangle$ which is a subsequence of $\langle a_n\rangle$ and is monotone nondecreasing. QED

Lemma 1.9. Let $\langle a_n \rangle$ be a sequence with the following property. There exists a nonnegative integer N such that no element of the sequence $\langle a_{n,N} \rangle$ is a lower bound for the sequence $\langle a_{n,N} \rangle$. Then $\langle a_n \rangle$ has a monotone decreasing subsequence.

Proof. Let $b_1 = a_{N+1}$. By hypothesis, b_1 is not a lower bound for the sequence $\langle a_{n,N} \rangle$. This means that there is some term of the sequence $\langle a_{n,N} \rangle$ whose value is less than b_1; hence, there exists a smallest positive integer k_2 such that $a_{N+k_2} < b$. Define $b_2 = a_{N+k_2}$. However, b_2 is not a lower bound for the sequence $\langle a_{n,N} \rangle$, so there exists a smallest positive integer $k_3 > k_2$ such that $a_{N+k_3} < b_2$. Define $b_3 = a_{N+k_3}$. Then $b_1 > b_2 > b_3$. In general, suppose we have determined j terms $b_1 > b_2 > b_3 > \ldots > b_j$, where for each integer i such that $1 \leq i \leq j$, $b_i = a_{N+k_i}$. Then, since b_j is not a lower bound for $\langle a_{n,N} \rangle$, there exists a smallest positive integer $k_{j+1} > k_j$ such that $a_{N+k_{j+1}} < b_j$. Define $b_{j+1} = a_{N+k_{j+1}}$. We thus get $j + 1$ terms $b_1 > b_2 > b_3 > \ldots > b_j > b_{j+1}$, where for each integer i such that $1 \leq i \leq j + 1$, $b_i = a_{N+k_i}$. Therefore we have defined inductively the entire sequence $\langle b_n \rangle$ which is a subsequence of $\langle a_n \rangle$ and is monotone decreasing. QED

We are now prepared to state and prove the following important theorem.

Theorem 1.10. Every sequence of real numbers has a monotone subsequence.

Proof. Let $\langle a_n \rangle$ be any sequence of real numbers. Then exactly one of the following statements is true:

(i) For every nonnegative integer N, there exists an element of the sequence $\langle a_{n,N} \rangle$ which is a lower bound of the sequence $\langle a_{n,N} \rangle$.

(ii) There exists a nonnegative integer N such that no element of the sequence $\langle a_{n,N} \rangle$ is a lower bound for the sequence $\langle a_{n,N} \rangle$.

Now, if (i) is true, $\langle a_n \rangle$ contains a monotone nondecreasing subsequence by Lemma 1.8, and if (ii) is true, then $\langle a_n \rangle$ contains a monotone decreasing subsequence by Lemma 1.9. Thus, in either case, $\langle a_n \rangle$ has a monotone subsequence. QED

PROBLEMS

1. Show that if $\langle x_n \rangle$ is a sequence which is simultaneously monotone nonincreasing and monotone nondecreasing, then it must be a constant sequence; that is, all terms are equal.

2. Prove Theorem 1.7 (*b*), (*c*), and (*d*).
3. Prove that every monotone nondecreasing sequence has a lower bound.
4. Prove that every monotone nonincreasing sequence has an upper bound.
5. Prove that a monotone nondecreasing sequence is bounded iff it has an upper bound.
6. Prove that a monotone nonincreasing sequence is bounded iff it has a lower bound.
7. Let $\langle x_n \rangle$ be a sequence, and define the sequence $\langle y_n \rangle$ as follows: For every positive integer n, $y_n = -x_n$. Then prove the following:
   (*a*) $\langle x_n \rangle$ is monotone nonincreasing iff $\langle y_n \rangle$ is monotone nondecreasing.
   (*b*) $\langle x_n \rangle$ is monotone nondecreasing iff $\langle y_n \rangle$ is monotone nonincreasing.
   (*c*) $\langle x_n \rangle$ is monotone decreasing iff $\langle y_n \rangle$ is monotone increasing.
   (*d*) $\langle x_n \rangle$ is monotone increasing iff $\langle y_n \rangle$ is monotone decreasing.
8. Prove Theorem 1.2.
9. Prove Theorem 1.3.

2. *Uniformly isolated sequences*

Consider again the sequence $I_o = \langle 1, 2, 3, 4, \ldots \rangle$. We see that this sequence has the property that each pair of distinct terms i and j satisfies the inequality $|i - j| \geq 1$. As a matter of fact, if r is any real number such that $0 < r < 1$, and if $i, j \in I_o$ with $i \neq j$, then $|i - j| > r$. For this reason, we say that the sequence I_o is uniformly isolated.

Definition 2.1. A sequence $\langle x_n \rangle$ is *uniformly isolated* iff there exists a real number $r > 0$ such that $|x_i - x_j| > r$ for every pair of positive integers i, j with $i \neq j$.

According to Definition 2.1, every uniformly isolated sequence must be a sequence of distinct terms. This fact might tempt us to hope that every sequence of distinct terms is uniformly isolated. The following example eliminates any such hope.

Example 2.2. Let $\langle x_n \rangle$ be the sequence $\langle 1, \frac{1}{2}, \frac{1}{3}, \ldots, 1/n, \ldots \rangle$, which is defined formally as follows: For each $n \in I_o$, $f(n) = x_n = 1/n$. This is obviously a sequence of distinct terms. To prove that it is not uniformly isolated, let $r > 0$ be any positive number. By the Archimedean property, there exists a positive integer n such that $nr > 1$; consequently $n(n + 1)r > 1$. Hence $|x_n - x_{n+1}| = 1/n - 1/(n + 1) = 1/[n(n + 1)] < r$, and $\langle x_n \rangle$ is not uniformly isolated.

Theorem 2.3. If $\langle a_i \rangle$ is uniformly isolated and if $\langle b_i \rangle$ is a subsequence of $\langle a_i \rangle$, then $\langle b_i \rangle$ is uniformly isolated.

Proof. We are given that $\langle a_i \rangle$ is uniformly isolated. Therefore, there exists $r > 0$ such that

$$\text{(i)} \qquad |a_i - a_j| > r \quad \text{if} \quad i \neq j$$

Now let b_k and b_m, where $k \neq m$, be any two points of $\langle b_i \rangle$. We will have succeeded in proving the theorem if we can show that $|b_k - b_m| > r$. Since $\langle b_i \rangle$ is given as a subsequence of $\langle a_i \rangle$, there exists a subsequence $\langle n_i \rangle$ of I_o such that $b_i = a_{n_i}$ for each i. Then, in particular,

$$\text{(ii)} \qquad b_k = a_{n_k} \qquad \text{and} \quad b_m = a_{n_m}$$

But $\langle n_i \rangle$ is a sequence of distinct terms, since it is a subsequence of I_o, so $k \neq m$ implies that $n_k \neq n_m$, and hence by (i) we have $|a_{n_k} - a_{n_m}| > r$. Finally, using (ii), we obtain $|b_k - b_m| > r$. Therefore $\langle b_i \rangle$ is uniformly isolated. QED

Theorem 2.3 tells us that every subsequence of a uniformly isolated sequence is uniformly isolated. It is also possible, however, for a nonuniformly isolated sequence to have a uniformly isolated subsequence, as shown by the following example.

Example 2.4. Let $\langle a_n \rangle$ be the sequence $\langle 1, 2, 1, 4, 1, 6, 1, 8, \ldots \rangle$, which is defined formally as follows: For each positive integer n,

$$f(n) = a_n = \begin{cases} 1 & \text{if } n \text{ is odd} \\ n & \text{if } n \text{ is even,} \end{cases} \qquad \text{or } f(n) = \frac{n+1}{2} + (-1)^n \cdot \frac{n-1}{2}$$

The sequence $\langle a_n \rangle$ is obviously not uniformly isolated, since it is not even a sequence of distinct terms. However, the subsequence $\langle 2, 4, 6, 8, \ldots \rangle$ consisting of the even terms is a subsequence of I_o, and hence is uniformly isolated by Theorem 2.3.

In order to prove the principal theorem of this section, we make use of the following lemma.

Lemma 2.5. Every unbounded subset H of R_1 contains a monotone sequence of points which is uniformly isolated.

****Proof.*** We suppose that H has no upper bound, and leave the case where H has no lower bound as an exercise.

For each positive integer i, define

$$H_i = H \cap \{x : i - 1 \leq x \leq i\}$$

This gives us the following sequence of sets:

$$\langle H_1, H_2, H_3, \ldots \rangle$$

Since H has no upper bound, not all of the sets of this sequence can be empty. By the Well-Ordering Axiom for I_o, we know that there exists a smallest positive integer i_1 such that H_{i_1} is nonempty.

Consider the sequence of sets

$$\langle H_{i_1+2}, H_{i_1+3}, H_{i_1+4}, \ldots, H_{i_1+i}, \ldots \rangle$$

Since H has no upper bound, not all these sets can be empty. By the Well-Ordering Axiom for I_o, there must exist a smallest positive integer j_2 such that $H_{i_1+j_2}$ is nonempty. We define $i_2 = i_1 + j_2$. Note that $H_{i_2} \neq \phi$ and $i_2 - i_1 \geq 2$.

Suppose that the set H_{i_k} has been defined to satisfy $H_{i_k} \neq \phi$ and $i_k - i_{k-1} \geq 2$. Consider the sequence of sets

$$\langle H_{i_k+2}, H_{i_k+3}, H_{i_k+4}, \ldots, H_{i_k+i}, \ldots \rangle$$

Since H has no upper bound, not all these sets can be empty. By the Well-Ordering Axiom for I_o, there must exist a smallest positive integer j_{k+1} such that $H_{i_k+j_{k+1}}$ is nonempty. Define $i_{k+1} = i_k + j_{k+1}$. Note that $H_{i_{k+1}} \neq \phi$ and $i_{k+1} - i_k \geq 2$. We have thus defined by mathematical induction a sequence of sets

$$\langle H_{i_1}, H_{i_2}, H_{i_3}, \ldots, H_{i_j}, \ldots \rangle$$

such that

(1) $H_{i_j} \neq \phi$ for every integer j.

(2) $i_{j+1} - i_j \geq 2$ for every integer j.

It should be noted from (2) that the following condition also holds.

(3) If $j < k$, then $i_k - i_j \geq 2$.

By the Axiom of Choice for Sequences, there exists a sequence $\langle p_n \rangle$ of points such that $p_n \in H_{i_n}$ for every positive integer n. Furthermore, it follows from the definition of H_{i_j} and condition (3) that if m and k are positive integers and $m < k$, then

$$i_m - 1 \leq p_m \leq i_m < i_m + 1 \leq i_k - 1 \leq p_k \leq i_k$$

Consequently, if $m < k$, then $p_k - p_m > 1$. Therefore the sequence $\langle p_n \rangle$ is both monotone increasing and uniformly isolated. QED

Theorem 2.6. A set of real numbers is unbounded iff it contains a monotone sequence which is uniformly isolated.

Proof. Half the theorem is already proved; that is, if a set is unbounded, it contains a monotone sequence which is uniformly isolated by Lemma 2.5. To complete the proof, we must show that if a set contains a monotone sequence which is uniformly isolated, the set is unbounded. Thus, let A be a set, and let $\langle x_n \rangle$ be a sequence of points in A such that $\langle x_n \rangle$ is monotone and uniformly isolated. Since $\langle x_n \rangle$ is uniformly isolated, there exists a positive number r such that $|x_i - x_j| > r$ for $i \neq j$. Then, in particular, $|x_{n+1} - x_n| > r$ for every n. Also $\langle x_n \rangle$ must be strictly monotone. Let as assume that $\langle x_n \rangle$ is monotone increasing (that is, $x_{n+1} > x_n$ for every n) and leave the case where the sequence is monotone decreasing as an exercise. Thus $x_{n+1} - x_n > 0$; so $|x_{n+1} - x_n| = x_{n+1} - x_n > r$ for every n, or $x_{n+1} > x_n + r$ for every n. We thus get the following chain of inequalities:

(1)	$x_2 > x_1 + r$	
(2)	$x_3 > x_2 + r > x_1 + 2r$	by (1)
(3)	$x_4 > x_3 + r > x_1 + 3r$	by (2)

and, in general, if k is any positive integer, then

$$x_{k+1} > x_k + r > x_1 + kr$$

Now, if M is any positive number, no matter how large, we can choose the positive integer k so large that $x_1 + kr > M$, by the Archimedean property. But then M cannot be an upper bound for A, since $x_{k+1} > M$, and $x_{k+1} \in A$. Hence, A has no upper bound, and is therefore unbounded. QED

PROBLEMS

10. Prove Lemma 2.5 for the case in which H has no lower bound.
11. State precisely, in your best mathematical language, what is meant by the statement "The sequence $\langle x_n \rangle$ is not uniformly isolated."
12. Let $\langle x_n \rangle$ be a sequence and define the sequence $\langle y_n \rangle$ as follows: For every positive integer n, $y_n = -x_n$. Then prove the following:

(*a*) $\langle x_n \rangle$ is uniformly isolated iff $\langle y_n \rangle$ is uniformly isolated.
(*b*) $\langle x_n \rangle$ is bounded below iff $\langle y_n \rangle$ is bounded above.
(*c*) $\langle x_n \rangle$ is bounded above iff $\langle y_n \rangle$ is bounded below.
(*d*) $\langle x_n \rangle$ is bounded iff $\langle y_n \rangle$ is bounded.

13. Prove: Every monotone, uniformly isolated sequence is unbounded.
14. Prove: Every set which contains a uniformly isolated sequence of points is unbounded.
15. Prove: Every uniformly isolated sequence is unbounded.

16. Prove that if a monotone sequence is uniformly isolated, then it is strictly monotone.

17. Prove: A sequence is unbounded iff it has a uniformly isolated subsequence which is monotone.

18. Prove the second half of Theorem 2.6 by making careful use of Problems 7 and 12.

19. Determine in each case whether or not the given sequence is uniformly isolated.

(*a*) $\left\langle \sin \frac{1}{n} \right\rangle$ (*b*) $\langle \sin n \rangle$ (*c*) $\left\langle \cot \frac{1}{n} \right\rangle$ (*d*) $\langle \log n \rangle$

III

Countable, connected, open, and closed sets

1. Countable sets

In our discussion of sets and sequences in Chapter I we pointed out that although all sequences are sets, not all sets are sequences. We saw, in fact, that in order to make a set A into a sequence, we must be able to find a mapping from I_o onto A; that is, we must be able to index the set A by the set of positive integers with their natural ordering. Now, since we use the positive integers for counting, it seems reasonable to assume that any set which can be indexed by the set of positive integers can be counted, and is thus what we call a countable set. For convenience, we define the empty set ϕ to be countable.

Definition 1.1. The empty set ϕ is *countable*. A nonempty set H is *countable* iff there exists a mapping f of I_o onto H.

Theorem 1.2. The set I is countable.

Proof. See Example 2.15, Chapter I.

In order to prove certain theorems about countable sets, we now find it convenient to define the concept of the composite mapping of two given mappings.

Definition 1.3. Let A, B, C be nonempty sets, and f, g mappings such that $f\colon A \to B$ and $g\colon B \to C$. Then the *composite* *mapping* of g with f is a mapping $h: A \rightarrow C$, where h is defined by $h(x) = g[f(x)]$ for every $x \in A$, and we write $h = gf$.

Example 1.4. Let $A = \{a, b, c, d, e, x, y, z\}$; $B = \{i, j, k, l, m\}$; $C = \{1, 3, 5, 7\}$. Define the mapping $f: A \xrightarrow{\text{onto}} B$ as follows:

$$\begin{aligned} f(a) &= k & f(e) &= l \\ f(b) &= m & f(x) &= j \\ f(c) &= i & f(y) &= j \\ f(d) &= j & f(z) &= j \end{aligned}$$

Now, define the mapping $g: B \xrightarrow{\text{onto}} C$ as follows:

$$\begin{aligned} g(i) &= 3 & g(l) &= 3 \\ g(j) &= 5 & g(m) &= 1 \\ g(k) &= 7 \end{aligned}$$

Let h be the composite mapping gf. Then we obtain

$$\begin{aligned} h(a) &= gf(a) = g[f(a)] = g(k) = 7 \\ h(b) &= gf(b) = g[f(b)] = g(m) = 1 \\ h(c) &= gf(c) = g[f(c)] = g(i) = 3 \\ h(d) &= gf(d) = g[f(d)] = g(j) = 5 \\ h(e) &= gf(e) = g[f(e)] = g(l) = 3 \\ h(x) &= gf(x) = g[f(x)] = g(j) = 5 \\ h(y) &= gf(y) = g[f(y)] = g(j) = 5 \\ h(z) &= gf(z) = g[f(z)] = g(j) = 5 \end{aligned}$$

Thus we see that the composite mapping h is a mapping from A onto C; that is, $h = gf: A \xrightarrow{\text{onto}} C$.

Example 1.5. Let $B = \{1, 3, 5, 7, 9, 11, \ldots\}$, and define $f: I_o \xrightarrow{\text{onto}} B$ by $f(x) = 2x - 1$ for every $x \in I_o$. Let $C = \{2, 4, 6, 8, 10, \ldots\}$, and define $g: B \rightarrow C$ by $g(y) = 2y$ for every $y \in B$. Note that g is a mapping of B into C which is not an onto mapping. The composite mapping $h: I_o \rightarrow C$ is defined by $h = gf$; that is, $h(x) = g[f(x)]$ for every $x \in I_o$. In this particular example, $h(x) = g(2x - 1) = 4x - 2$ for every $x \in I_o$. Thus h maps I_o onto the set of all numbers, each of which is twice an odd positive integer. If we let H denote this set, then $H = \{2, 6, 10, 14, \ldots\}$, and the composite mapping h maps I_o onto H. However, h maps I_o into C, not onto C, since the set H here is a subset of C but is not equal to C.

Note that in Example 1.4 the composite mapping is onto C, whereas in Example 1.5 the composite mapping is into (but not onto) C. The next theorem presents a case where we can be sure that a composite mapping is onto.

Theorem 1.6. If $f: A \xrightarrow{\text{onto}} B$ and $g: B \xrightarrow{\text{onto}} C$, then the composite mapping gf is a mapping from A onto C; that is, $gf: A \xrightarrow{\text{onto}} C$.

Proof. We know that gf is a composite mapping from A into C by Definition 1.3. Thus we need only prove that it is onto. We do this by showing that every element in C is the image under the composite mapping gf of some point in A. Thus let p be any point in C. Since g is onto, there exists some point, say q, in B such that $g(q) = p$. But f is also onto, so there exists some point, say r, in A such that $f(r) = q$. Therefore

$$p = g(q) = g[f(r)] = gf(r)$$ QED

Theorem 1.7. Given nonempty sets A, B, and suppose there exists a mapping $f: A \xrightarrow{\text{onto}} B$. If A is countable, then B is countable.

Proof. Since A is countable, there exists a mapping $g: I_o \xrightarrow{\text{onto}} A$. We are given the mapping $f: A \xrightarrow{\text{onto}} B$. Thus, denoting the composite mapping fg by h, we have $h: I_o \xrightarrow{\text{onto}} B$ by Theorem 1.6. Therefore B is countable. QED

Lemma 1.8. If B is any nonempty subset of a set A, then there exists a mapping $f: A \xrightarrow{\text{onto}} B$.

Proof. We are given that $B \neq \phi$, so B must contain at least one point. Thus, let b be any point of B. Then we may define f as follows: For any $x \in A$,

$$f(x) = \begin{cases} x & \text{if } x \in B \\ b & \text{if } x \notin B \end{cases}$$

Clearly, f is a mapping from A onto B. QED

Example 1.9. Let $A = \{a, b, c, d\}$ and $B = \{a, c\}$. We can define $f: A \xrightarrow{\text{onto}} B$ as follows:

$$f(a) = a \qquad f(c) = c$$
$$f(b) = a \qquad f(d) = a$$

Of course, we could just as well have mapped b and d onto the point c.

Theorem 1.10. Every subset of a countable set is countable.

The proof is immediate, from Theorem 1.7 and Lemma 1.8.

Throughout our work, we are particularly interested in the problem of determining those properties of sets or sequences which are preserved under certain types of mappings. Theorem 1.7 tells us that if the domain of an onto mapping is countable, its range is also countable; or, more concisely, onto mappings preserve countability.

Definition 1.11. A nonempty subset H of I is a *finite set of positive integers* iff H is bounded. An arbitrary nonempty set A is *finite* iff there exists a finite set H of positive integers and a mapping $f: H \xrightarrow{\text{onto}} A$. The set ϕ is said to be finite.

Theorem 1.12. Every finite set is countable.

Proof. Let A be a finite set. If $A = \phi$, then A is countable. Suppose $A \neq \phi$. Then there exists a finite set H of positive integers and a mapping $f: H \xrightarrow{\text{onto}} A$. But $H \subset I$ and hence is countable by Theorems 1.2 and 1.10. Therefore A is countable by Theorem 1.7.

QED

We shall find it convenient to have at our disposal the concepts of one-to-one mappings and of sets consisting of exactly n elements.

Definition 1.13. Let f be a mapping of a set H into a set K. Then f is *one-to-one* iff given any pair of points x and y of H such that $x \neq y$, we have $f(x) \neq f(y)$.

Definition 1.14. We use the symbol I_n to denote the set $\{1, 2, 3, \ldots, n\}$, where n is a positive integer.

Theorem 1.15. For each positive integer n, the set I_n has the following properties:

(*a*) I_n is a bounded set of positive integers.
(*b*) I_n is a finite set of positive integers.
(*c*) I_n consists of exactly n elements.

The proof of Theorem 1.15 is left as an exercise. For the proof of (*c*), make use of the following definition.

Definition 1.16. A set H is said to *consist of exactly n elements* iff there exists a one-to-one mapping $f: I_n \xrightarrow{\text{onto}} H$.

Theorem 1.17. If a set H consists of exactly n elements, where n is a positive integer, then H is a finite set.

The proof is left as an exercise.

Definition 1.18. An *infinite set* is a set which is not finite. A *countably infinite set* is an infinite set which is countable, and an *uncountably infinite set* is a set which is not countable.

We shall prove that a set H is infinite iff it contains a sequence of distinct points. As a step in this direction, we show that given any positive integer n, the set H contains a subset consisting of exactly n elements.

Lemma 1.19. Let H be any infinite set, and n any positive integer. Then there exists a subset B of H consisting of exactly n elements.

Proof. We prove the lemma by mathematical induction. First of all, we see that H is not empty, since H is not finite. Let p be an element of H. Define a mapping $f: I_1 \xrightarrow{\text{onto}} \{p\}$ by $f(1) = p$. It follows at once that the set $\{p\}$ is a subset of H consisting of exactly one element. Therefore the lemma is true for $n = 1$.

We complete our inductive argument by showing that if the lemma is true when $n = k$, it must also be true when $n = k + 1$. Suppose that the lemma is true when $n = k$. Then there exists a subset B_k of H such that B_k consists of exactly k elements. Accordingly, there must exist a one-to-one mapping $g: I_k \xrightarrow{\text{onto}} B_k$. We see that $B_k \neq H$ since B_k is finite and H is not finite. Consequently, since $B_k \subset H$, there must exist a point p in H such that p is not in B_k. Let p be any such point. Define the set $B_{k+1} = B_k \cup \{p\}$. We now define a mapping $h: I_{k+1} \xrightarrow{\text{onto}} B_{k+1}$ as follows:

$$h(x) = \begin{cases} g(x) & \text{for every } x \in I_k \\ p & \text{for } x = k + 1 \end{cases}$$

It is left for the student to show that h is both one-to-one and onto. It follows from these facts that B_{k+1} is a set consisting of exactly $k + 1$ elements. QED

Lemma 1.20. If H is finite, then H does not contain a sequence of distinct points.

Proof. Suppose that H contains a sequence $\langle b_i \rangle$ of distinct points. Since H is finite, there exists a bounded set B of positive integers and a mapping $f: B \xrightarrow{\text{onto}} H$. The fact that B is bounded tells us that there is a positive integer n which is an upper bound for B. Consequently, $B \subset I_n$. By Lemma 1.8, there exists a mapping $g: I_n \xrightarrow{\text{onto}} B$. By Theorem 1.6, the composite mapping $h = fg$ is a mapping of I_n onto H. We write this mapping

$$h: I_n \xrightarrow{\text{onto}} H$$

Consider the $n + 1$ distinct points $b_1, b_2, \ldots, b_{n+1}$. We know that for each $i = 1, 2, 3, \ldots, n + 1$, there exists a point a_i in I_n such that $h(a_i) = b_i$. If $i \neq j$, then $b_i \neq b_j$; that is, $h(a_i) \neq h(a_j)$. Therefore $a_i \neq a_j$. This proves that I_n contains the $n + 1$ distinct elements $a_1, a_2, \ldots, a_{n+1}$, and this contradicts Theorem 1.15(c). Since we have obtained a contradiction, our original assumption must be false. This completes the proof. QED

The contrapositive of Lemma 1.20 is worth stating in its own right.

Lemma 1.21. If H contains a sequence of distinct points, then H is infinite.

We are now ready for our principal theorem on infinite sets.

Theorem 1.22. A set H is infinite iff H contains a sequence of distinct points.

****Proof.*** If H contains a sequence of distinct points, we know that H is infinite by Lemma 1.21. Thus, we need only prove that if H is any infinite set, H contains a sequence of distinct points. Let H be any infinite set. Let n be any positive integer. By Lemma 1.19, there exists a subset B of H consisting of exactly n elements. Let $\mathfrak{M}_n$ be the collection of all subsets of H, each of which consists of exactly n elements. By our previous remark, for every positive integer n, the set $\mathfrak{M}_n$ is nonempty. Consider the sequence of nonempty sets $\langle \mathfrak{M}_n \rangle$. By the Axiom of Choice for Sequences, we see that there exists a sequence $\langle A_n \rangle$ such that for each positive integer n, $A_n \in \mathfrak{M}_n$. This means precisely that A_n is a subset of H consisting of exactly n elements. Consequently, there exists a one-to-one mapping $f: I_n \xrightarrow{\text{onto}} A_n$. Define $\mathcal{H}_n$ as the set of all one-to-one mappings of I_n onto A_n. Note that for each positive integer n, $\mathcal{H}_n$ is not empty. Consider the sequence $\langle \mathcal{H}_n \rangle$ of nonempty sets. By the Axiom of Choice for Sequences, there exists a sequence of mappings $\langle f_n \rangle$ such that $f_n \in \mathcal{H}_n$ for every positive integer n.

We are now prepared to define a sequence of distinct points $\langle c_i \rangle$ of H. Consider the mapping $f_1: I_1 \xrightarrow{\text{onto}} A_1$, and define $c_1 = f_1(1)$. Note that $A_1 = \{c_1\}$. We next look at the mapping $f_2: I_2 \xrightarrow{\text{onto}} A_2$, and observe that at least one of the elements $f_2(1)$ and $f_2(2)$ is different from c_1. Let i_1 be the smallest positive integer such that $f_2(i_1) \neq c_1$. Define $c_2 = f_2(i_1)$. Note that $c_2 \neq c_1$. We proceed by mathematical induction. Suppose that we have found k distinct elements $c_1, c_2, c_3, \ldots, c_k$ such that c_i is an element of A_i for $i = 1, 2, 3, \ldots, k$. Consider the mapping $f_{k+1}: I_{k+1} \xrightarrow{\text{onto}} A_{k+1}$. The set A_{k+1} consists of exactly $k + 1$ elements. Thus at least one of the elements

$$f_{k+1}(1), f_{k+1}(2), f_{k+1}(3), \ldots, f_{k+1}(k + 1)$$

is distinct from all the elements $c_1, c_2, c_3, \ldots, c_k$. Let i_k be the smallest positive integer such that $f_{k+1}(i_k) \neq c_i$ for $i = 1, 2, 3, \ldots, k$. Define $c_{k+1} = f_{k+1}(i_k)$. We have thus defined inductively the sequence $\langle c_i \rangle$ of distinct elements of H. QED

We have already dealt with several countably infinite sets, notably the set I of positive integers, as well as a few infinite subsets of I (for example, the set of all positive even integers). It should be clear at this point that those infinite sets which can be made into sequences are precisely the countably infinite sets. It should also be noted that the class of countable sets includes the empty set, all finite sets, and all those sets which are countably infinite. We now proceed to exhibit a few additional infinite sets which are countable.

We recall that two integers are *relatively prime* iff they have no common factor greater than one. The integers 3 and 5 in the following lemma could be any pair of positive integers which are relatively prime.

Lemma 1.23. Let K be the set of all positive integers x such that x can be written in the form $x = 3^h 5^k$, where h and k are nonnegative integers. Then K is countable.

Proof. This follows at once from Theorems 1.2 and 1.10, since $K \subset I$. QED

Theorem 1.24. Let B be the set of all pairs (h, k), where h and k are nonnegative integers. Then B is countable.

Proof. Let K be the set defined in Lemma 1.23, and define the mapping f as follows: For each $x = 3^h 5^k$ in K, $f(x) = (h, k)$. It is easily seen that f maps K onto B. Therefore B is countable by Lemma 1.23 and Theorem 1.7. QED

Recall that a rational number is a number which can be expressed as the quotient of two integers; that is, the number r is rational iff $r = p/q$, where p and q are integers with $q \neq 0$. Such numbers (here we include the integers as fractions with appropriate denominators) were called *common fractions* in algebra. Recall also that two rational numbers a/b and c/d are equal iff $ad = bc$. Fractions satisfying this condition were called *equivalent fractions* in algebra. For example, $1/2 = 2/4 = 3/6 = 4/8 = \ldots$, and $2 = 2/1 = 4/2 = 6/3 = \ldots$. We thus see that each rational number can be represented by an infinite number of equivalent fractions. Without pursuing this discussion further, it suffices for our purposes here to recognize the fact that the set of rational numbers is a subset of the set of all common fractions. We then have the following important theorem.

Theorem 1.25. The set of all positive rational numbers is countable.

Proof. Let K be the set of all pairs (h, k), where h and k are positive integers. Then K is a subset of the countable set B of Theorem 1.24, and hence K is countable by Theorem 1.10. Let C be the set consisting of all fractions of the form h/k, where h and k are positive integers. Now we define the mapping f from K to C as follows: For every $x = (h, k)$ in K, $f(x) = h/k$. Clearly, f is an onto mapping, so C is countable by Theorem 1.7. But the set of positive rational numbers is a subset of C and is therefore countable by Theorem 1.10. QED

It is interesting to see how the set C in Theorem 1.25 can actually be counted. We arrange the fractions in rows and columns, where all the fractions in a given row have the same numerator, and all the fractions in a given column have the same denominator. We thus get the array shown in Figure 1. Note that the fraction h/k is the entry in row h and column k. It is thus evident that every fraction h/k, where h and k are positive integers, must be included in the array. We now count the terms in the order indicated by the arrowed path, where the first term is $1/1$, the second $1/2$, the third $2/1$, the fourth $3/1$, and so on.

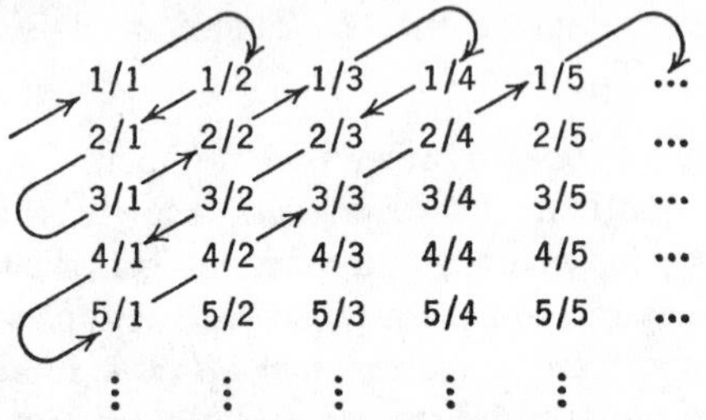

Figure 1

In order to prove further theorems regarding countable sets, we find it convenient to introduce the concepts of union and intersection for arbitrary collections of sets. These concepts are brought in by means of the notion of an index mapping. In Definition 2.10 of Chapter I, we used a set H to index a set K of points in the universe U. We now want to consider the problem of using a set H to index a collection of subsets of U. In other words, the members of the set to be indexed are sets rather than points in U. To avoid confusion, we generally use a different notation for such a collection of sets, designating it by a script letter such as $\mathcal{K}$. Consider the following example which illustrates the technique.

Example 1.26. Let the universe be $U = \{1, 2, 3, 4, 5, 6, 7, 8\}$, and let $H = \{1, 2, 3\}$. We know from our work in Chapter I that the set U has $2^8 = 256$ subsets. Let us denote by $\mathcal{U}$ the set of all such subsets of U. We thus see that $\mathcal{U}$ is a set consisting of exactly 256 elements, each of which is a subset of U. It happens that H is one of the 256 elements in $\mathcal{U}$ (since H is a subset of U), but this fact does not particularly concern us here. Let us now define the following subsets of U:

$$A = \{1, 2, 3, 5\}, \qquad B = \{2, 5, 6, 8\}, \qquad C = \{1, 4, 6, 7\}$$

Note that A, B, and C are distinct subsets of U, even though they have some points in common, and therefore they are distinct elements of the set $\mathcal{U}$. If we define $\mathcal{K} = \{A, B, C\}$, we see that $\mathcal{K}$ is a subset of $\mathcal{U}$ and consists of just 3 of the 256 elements in $\mathcal{U}$. Or, if we prefer, we may say that $\mathcal{K}$ is a set (or collection) of subsets of U. Regardless of which interpretation we use, the fact remains that $\mathcal{K}$ is a bona fide set, and we can therefore define a mapping from the set H onto the set $\mathcal{K}$. We thus define a mapping $f: H \xrightarrow{\text{onto}} \mathcal{K}$ as follows:

$$f(1) = A \qquad f(3) = C$$
$$f(2) = B$$

Now, let us rename the points of $\mathcal{K}$ in such a way as to associate with each point in H its corresponding image in $\mathcal{K}$. We define

$$K_1 = f(1) \qquad K_3 = f(3)$$
$$K_2 = f(2)$$

We may now write $\mathcal{K} = \{K_1, K_2, K_3\}$, where, of course, $K_1 = A$, $K_2 = B$, and $K_3 = C$. Writing $\mathcal{K}$ in the form $\mathcal{K} = \{K_1, K_2, K_3\}$, we say that the collection has been indexed by the set H under the mapping f.

This example illustrates the technique for indexing a collection of sets. We now generalize this concept in the following definition.

Definition 1.27. Let U be the universe, H a nonempty set, and $\mathcal{K}$ a collection of subsets of U. If $f: H \xrightarrow{\text{onto}} \mathcal{K}$, and if for each $\alpha \in H$ we define the image of α under f in $\mathcal{K}$ by $f(\alpha) = K_\alpha$, then we say that the collection $\mathcal{K} = \{K_\alpha\}$ has been *indexed by the set H* under the mapping f. The mapping f is called the *index mapping*, and the set H is called the *index set.*

We often bypass many of the details in Definition 1.27 by making a statement such as the following: "Let H be an index set, and for every $p \in H$, let A_p be a set." Here, of course, it is assumed that we have a universe U, a nonempty index set H, a collection $\mathcal{A}$ of subsets of U, and an index mapping f under which the collection $\mathcal{A} = \{A_p\}$ has been indexed by H. Note further that we have placed no restrictions on the index set H in the statement of Definition 1.27 (except that H be nonempty), so that H may be a finite set, or it may be either countably infinite or uncountably infinite. Thus, the collection $\{A_p\}$, where $p \in H$, may be finite, countably infinite, or uncountably infinite.

Also, in Chapter I, we defined the intersection of two sets (Definition 1.7) and the union of two sets (Definition 1.8). We then showed how these definitions may be extended to apply to any *finite* number of sets. With the aid of Definition 1.27, we can now extend the notions of union and intersection to an arbitrary number of sets.

Definition 1.28. Let U be the universe, H an index set, and for every $p \in H$, let A_p be a set. Then the *union of the collection of sets* A_p (written $\bigcup_{p \in H} A_p$) is the set consisting of all those points $x \in U$ such that $x \in A_p$ for at least one $p \in H$.

As might be expected, Definition 1.28 says that a point is in the union of a collection of sets iff it is a member of at least one set in the collection.

Definition 1.29. Let U be the universe, H an index set, and for every $p \in H$, let A_p be a set. Then the *intersection of the collection of sets* A_p (written $\bigcap_{p \in H} A_p$) is the set consisting of all those points $x \in U$ such that $x \in A_p$ for every $p \in H$.

Again as expected, Definition 1.29 says that a point is in the intersection of a collection of sets iff it is a member of every set in the collection.

The statements in the following theorem are easy consequences of the above definitions, and their proofs are left as exercises.

Theorem 1.30. Let H be an index set, and for every $p \in H$, let A_p be a set. Then the following statements are true:

(*a*) If $q \in H$, then $A_q \subset \bigcup_{p \in H} A_p$.

(*b*) If $q \in H$, then $\bigcap_{p \in H} A_p \subset A_q$.

If in Figure 1 we designate the set of fractions in the first row by A_1, the set in the second row by A_2, and so on, we see that the set C is the union of all the sets A_i, where i is a positive integer. Now, since each set A_i is countable, and since there are a countable number of such sets A_i, we may say that C is the union of a countable collection of countable sets. We have already shown that C is countable, so we may naturally wonder whether or not the union of every countable collection of countable sets is countable.

Theorem 1.31. The union of any countable collection of countable sets is countable.

Proof. For each positive integer h, let A_h be a countable set, and let J be the union of the sets A_h. We want to show that J is countable. If all the sets A_h are empty, then J is empty, and hence countable. Having disposed of this trivial case, we may now suppose that at least one of the sets is nonempty; in fact, we lose no generality by assuming that A_1 is nonempty, for otherwise we could relabel the sets to make it so. We use a double subscript notation, so that the kth term of the set A_h is designated by $a_{h,k}$. Although it is not essential to the proof, we may find it easier to visualize the set J if we arrange it in an array similar to that used in Figure 1. Our new array is shown in Figure 2.

$$\begin{array}{llllll}
A_1: & a_{1,1} & a_{1,2} & a_{1,3} & a_{1,4} & \cdots \\
A_2: & a_{2,1} & a_{2,2} & a_{2,3} & a_{2,4} & \cdots \\
A_3: & a_{3,1} & a_{3,2} & a_{3,3} & a_{3,4} & \cdots \\
A_4: & a_{4,1} & a_{4,2} & a_{4,3} & a_{4,4} & \cdots \\
\vdots & \vdots & \vdots & \vdots & \vdots &
\end{array}$$

Figure 2

Of course, we must remember that a finite set is countable, so that any particular row h (consisting of the members of the set A_h) may have only a finite number of entries. Since A_1 is not empty (by our previous assumption), it must contain at least one member, so we can be sure that $a_{1,1} \in A_1$. Let K be the set of all pairs (h, k), where h and k are positive integers. We have shown in the proof of Theorem 1.25 that the set K is countable, so we would like to find a mapping from K onto J. We define this mapping f as follows: For every $x = (h, k)$ in K,

$$f(x) = \begin{cases} a_{h,k} & \text{if } a_{h,k} \in A_h \\ a_{1,1} & \text{if } a_{h,k} \notin A_h \end{cases}$$

Clearly, f is a mapping from K onto J. Therefore J is countable, by Theorem 1.7. QED

Theorem 1.32. The set of negative rational numbers is countable.

Proof. We define the mapping f as follows: For every positive rational number x, $f(x) = -x$. This mapping f maps the positive rational numbers onto the negative rational numbers. Its domain is countable, by Theorem 1.25. Therefore its range is countable, by Theorem 1.7. QED

Theorem 1.33. The set of all rational numbers is countable.

Proof. The set of all rational numbers is the union of the following three sets:

(1) The set of positive rational numbers, which is countable by Theorem 1.25.

(2) The set of negative rational numbers, which is countable by Theorem 1.32.

(3) The set $\{0\}$, which is countable by Theorem 1.12.

Thus the set of all rational numbers is the union of a finite (and hence countable) collection of countable sets, and is therefore countable, by Theorem 1.31. QED

Corollary 1.34. The set of all integers is countable.

Proof. The set of all integers is a subset of the set of all rational numbers (consisting of those rational numbers p/q, where p is any integer and $q = 1$), and hence is countable, by Theorems 1.33 and 1.10. QED

We have now determined a large number of countably infinite sets, so many in fact that it may seem impossible that there could be such a thing as an uncountably infinite set. We shall show before long, however, that the set of real numbers is uncountable; in fact, even the set of real numbers between zero and one is uncountable.

The following theorem is the contrapositive of the definition of a countable set.

Theorem 1.35. A set K is not countable iff given any mapping f of I_o into K, then f is not an onto mapping.

Now let us consider decimals for a moment. We know that every real number can be written as a decimal; in particular, any real number between zero and one can be written in the form $0.a_1a_2a_3a_4 \ldots$, where each a_i is one of the digits 0 through 9. We saw in algebra that if the decimal terminates or repeats in regular cycles, then the number is rational; if the decimal is nonterminating and nonrepeating, then the number is irrational. Even for a terminating decimal, we can add zeros after the last nonzero digit and thus consider it as nonterminating; for example, we can write $\frac{1}{4} = 0.25$ as $0.250000\ldots$. Therefore we can consider every real number as having an infinite decimal representation. One small problem remains to be resolved. The decimal representation of a number is not necessarily unique. The decimals $0.9999\ldots$ and $1.0000\ldots$ represent the same number; also, the decimals $0.13249999\ldots$ and $0.13250000\ldots$ represent the same real number. To avoid this ambiguity, we adopt the convention that whenever such a choice of representation is possible, the decimal with the zeros is chosen as the representative of the number. With this convention, we now have a unique infinite decimal representation for every real number.

We are finally ready to prove the following theorem.

Theorem 1.36. The set A of all real numbers between zero and one is not countable.

Proof. We shall use Theorem 1.35. Let f be any mapping of I_o into A. We must prove that f is not an onto mapping. Every member of A has a unique infinite decimal representation, and hence the range of f may be displayed as shown in Figure 3.

$$\begin{aligned}
f(1) &= 0.a_{11}a_{12}a_{13}a_{14}\ldots\\
f(2) &= 0.a_{21}a_{22}a_{23}a_{24}\ldots\\
f(3) &= 0.a_{31}a_{32}a_{33}a_{34}\ldots\\
&\vdots\\
f(n) &= 0.a_{n1}a_{n2}a_{n3}a_{n4}\ldots\\
&\vdots
\end{aligned}$$

Figure 3

We now construct an infinite decimal b by working only with the main diagonal from upper left to lower right in Figure 3. We choose two digits, say 5 and 8, and look first at a_{11}. If $a_{11} \neq 5$, we then use 5 in the first decimal place of b; and if $a_{11} = 5$, we use 8 in the first decimal place of b. In either case, b differs from the first decimal in the list in the first decimal place. Next, we look at a_{22}. If

$a_{22} \neq 5$, we use 5 in the second decimal place of b; and if $a_{22} = 5$, we use 8 in the second place of b. In either case, b differs from the second decimal in the list in the second decimal place. We can continue this process down through the entire list. The nth decimal place of the nth decimal in the list is denoted by a_{nn}. If $a_{nn} \neq 5$, we enter 5 in the nth decimal place of b; if $a_{nn} = 5$, we enter 8 in the nth decimal place of b. In this way, we get the infinite decimal b such that for every positive integer n, b differs from the nth decimal in the list in the nth decimal place. Thus b is an infinite decimal which is not in the list. We conclude that for every $n \in I_o$, $f(n) \neq b$. Therefore f is not an onto mapping. Consequently by Theorem 1.35, the set A is uncountable. QED

PROBLEMS

1. Let A and B be sets and f a mapping such that $f: A \xrightarrow{\text{onto}} B$. Is the following statement true? "If B is uncountable, then A is uncountable." If so, can you prove it?
2. Prove that the collection of all pairs of rational numbers is countable.
3. Prove that the collection of all line segments of R_1 which have rational length and rational midpoints is countable.
4. Prove that every finite set of positive integers is a finite set.
5. If a set A is made into a sequence by means of a mapping $f: I_o \xrightarrow{\text{onto}} A$ which is one-to-one, then this sequence is a sequence of distinct terms. Prove this statement.
6. Prove Theorem 1.15.
7. Prove Theorem 1.17.
8. Fill in the missing steps at the end of the proof of Lemma 1.19.
9. In Theorem 1.24, give an example where the proof will not work if the two integers are not relatively prime.
10. Prove Theorem 1.30.

2. *Connected sets. Open intervals and open rays*

Among all the subsets of R_1, there are certain types of sets which play a vital role in analysis. The remainder of this chapter is devoted to a detailed study of these special sets. Bear in mind that our universe is the set R_1 of real numbers, or equivalently, the set of points on the real line, and that we shall use the terms "real number" and "point" interchangeably.

Definition 2.1. A subset H of R_1 is *connected* iff given any two points a, b of H, every point between a and b is a point of H. A point p is said to be *between* a and b iff either $a < p < b$ or $b < p < a$.

The negation of Definition 2.1 is also useful to us.

Theorem 2.2. A subset H of R_1 is not connected iff there exist two points a, b of H and a point p between a and b such that $p \notin H$.

As a consequence of Theorem 2.2, we see that in order for a set not to be connected, the set must contain at least two points. We thus have the following immediate result.

Lemma 2.3. The empty set ϕ is connected, and every set consisting of just one point $\{p\}$ is connected.

We now list in convenient catalogue form those subsets of R_1 which are connected.

Catalogue of connected sets in R_1

(1) The empty set ϕ.
(2) All sets consisting of just one point $\{p\}$.
(3) All sets of the form $\{x: a < x < b\}$
(4) All sets of the form $\{x: a \leq x \leq b\}$
(5) All sets of the form $\{x: a \leq x < b\}$
(6) All sets of the form $\{x: a < x \leq b\}$
where $a, b \in R_1$ with $a < b$.
(7) All sets of the form $\{x: a < x\}$
(8) All sets of the form $\{x: a \leq x\}$
(9) All sets of the form $\{x: x < a\}$
(10) All sets of the form $\{x: x \leq a\}$
where $a \in R_1$.
(11) The set R_1.

It is an easy matter to verify that all sets in the catalogue are connected. But can we be sure that the catalogue contains *all* the connected subsets of R_1? The answer to this question is "yes," as shown by the next theorem. Note the interesting technique in the proof of the theorem. We rely on another very important property of sets, that of boundedness, exploit the fact that a set must be either bounded or unbounded, and then exhaust all the various possibilities as separate cases.

Theorem 2.4. Every connected set in R_1 is in the above catalogue.

Proof. Let H be any connected set in R_1. Then, by Lemma 2.3, H could be the empty set ϕ, or a set consisting of just one point $\{p\}$. But these sets are included in the catalogue. Thus we may now assume that H contains at least two points. The set H must be either bounded or unbounded, so we treat these possibilities as separate cases.

Case 1. H is bounded. In this case, H must have both a supremum and an infimum, by Theorem 4.5 of Chapter I. Thus, let $a = \inf H$ and $b = \sup H$. Since H contains at least two points, we must have $a < b$. Let p be any point such that $a < p < b$. Then p is neither an upper bound nor a lower bound of H. Thus there exist $q \in H$, $r \in H$ such that $a \leq q < p < r \leq b$. Therefore, since H is connected, $p \in H$. This proves

$$\text{(i)} \qquad \{x: a < x < b\} \subset H$$

From the definitions of a and b, we have at once that

$$\text{(ii)} \qquad H \subset \{x: a \leq x \leq b\}$$

The only sets H satisfying both (i) and (ii) are the connected sets of types (3), (4), (5), and (6) in the catalogue. Therefore every bounded connected set is of one of these types.

Case 2. H is bounded above, but not bounded below. In this case, define $a = \sup H$. We leave it as an exercise to show that

$$\{x: x < a\} \subset H \subset \{x: x \leq a\}$$

Thus H must be a connected set of type (7) or (8).

Case 3. H is bounded below, but not bounded above. The proof is left as an exercise.

Case 4. H is neither bounded above nor bounded below. The proof is left as an exercise. QED

Connected sets of type (3) as listed in the Catalogue of Connected Sets play a fundamental role in the remainder of this book. For this reason, we give these sets a special name.

Definition 2.5. An *open interval* in R_1 is a set of the form $\{x: a < x < b\}$, where $a, b \in R_1$ with $a < b$. We normally denote this set as the open interval (a, b). The points a and b are called the *endpoints* of the open interval (a, b), and the positive number $b - a$ is called the *length* of the open interval (a, b).

A few properties of open intervals should be immediately apparent from this definition. First, we see that an open interval is a connected set. Second, every open interval (a, b) has finite length $b - a > 0$, or, put another way, the open interval (a, b) is bounded; in fact, $a = \inf (a, b)$ and $b = \sup (a, b)$. We note that the endpoints of an open interval do not belong to the set. Thus we can characterize an open interval as a bounded, connected, nonempty set which contains neither its supremum nor its infimum. We point out also that neither the empty set nor a set consisting of a single point is an open interval.

Definition 2.6. An *open ray* in R_1 is a set of the form $\{x:x < a\}$ or the form $\{x: a < x\}$, where $a \in R_1$. In either case, the point a is called the *endpoint* of the open ray.

Just as with open intervals, we see that open rays are connected sets (items (7) and (9) in the Catalogue of Connected Sets). However, unlike open intervals, open rays are not bounded, and hence do not have finite length. Observe that an open ray is not an open interval. On the other hand, every open ray is the union of a countable collection of open intervals, as we now prove.

Theorem 2.7. Every open ray A is the union of a countable collection of open intervals.

Proof. We prove only the case where A is of the form $\{x: a < x\}$, and leave the other case as an exercise. For each positive integer n, define the open interval $J_n = (a, a + n)$. We leave it for the student to show that $A = \bigcup_{n \in I_0} J_n$. This is easily accomplished using the Archimedean property of R_1. QED

Corollary 2.8. R_1 is the union of a countable collection of open intervals.

The proof is left as an exercise.

PROBLEMS

11. Complete the proof of Case 2 of Theorem 2.4.
12. Prove Case 3 of Theorem 2.4.
13. Prove Case 4 of Theorem 2.4.
14. Given an index set H, suppose that for each $p \in H$, A_p is a connected

set. Suppose further that if p and q are any two elements of H, then $A_p \cap A_q \neq \phi$. Prove that $\bigcup_{p \in H} A_p$ is connected.

15. Given an index set H, suppose that for each $p \in H$, A_p is a connected set. Prove that $\bigcap_{p \in H} A_p$ is connected.

16. Complete the proof of Theorem 2.7.

17. Prove Corollary 2.8.

3. Open sets and components

We now use open intervals to define the more general (and much more important) concept of an open set in R_1.

Definition 3.1. A set G of real numbers is an *open set* in R_1 iff for each point $p \in G$, there exists an open interval (a, b) such that $p \in (a, b) \subset G$.

It is extremely important that we understand precisely what is meant by Definition 3.1. If we want to show that a given nonempty (the empty set will be disposed of shortly) set G is open, we first choose an arbitrary point $p \in G$. Then, we must be able to find an open interval (a, b) which contains p and which is contained in G; that is, every point of the open interval (a, b), including the point p, must be a member of the set G. It is worth noting that we need to find only one such open interval, and its length may be very small. However, having found one, we have many more such open intervals at our disposal; for, if the open interval (a, b) meets the requirements of Definition 3.1, and if c and d are any real numbers such that $a < c < p < d < b$, then the open interval (c, d) also meets the requirements of Definition 3.1. With this discussion in mind, we can now state the negation of Definition 3.1.

Theorem 3.2. A set G is not an open set iff there exists a point $p \in G$ such that every open interval which contains p also contains at least one point $q \notin G$.

Theorem 3.3. The empty set ϕ is an open set.

The contrapositive of Theorem 3.3 is: If a set is not open, then it is not empty. This is immediate from Theorem 3.2.

The next theorem has several useful corollaries.

Theorem 3.4. If a set A is the union of a collection of open intervals $\{J_p: p \in H\}$, where H is an index set, then A is open.

Proof. Let $x \in A$. Then there exists $p_o \in H$ such that

$$x \in J_{p_o} \subset A.$$

Therefore A is open, by Definition 2.1. QED

Corollary 3.5. Every open interval is an open set.

Corollary 3.6. Every open ray is an open set.

Proof. See Theorem 2.7.

Corollary 3.7. R_1 is an open set.

Proof. See Corollary 2.8.

Theorem 3.8. The intersection of two open intervals is either the empty set or an open interval.

Proof. Let (a, b) and (c, d) be two open intervals. If (a, b) and (c, d) have no point in common, their intersection is the empty set. Thus let us assume that they have at least one point in common. Since $(a, b) \cap (c, d) \neq \phi$, we must have $c < b$. Define $r = \sup \{a, c\}$, and $s = \inf \{b, d\}$. Then it can be shown that $(a, b) \cap (c, d) = (r, s)$. This is left as an exercise. QED

Theorem 3.9. An open set contains neither its supremum nor its infimum.

Proof. Let G be any open set in R_1, and let p be any point in G. Let (a, b) be any open interval such that $p \in (a, b) \subset G$. Then there exist points $c, d \in G$ such that $a < c < p < d < b$. Therefore p is neither the supremum nor the infimum of G. QED

Theorem 3.10. The set $\{p\}$ consisting of just one point in R_1 is not an open set.

Proof. The set $\{p\}$ contains both its supremum and infimum (both are equal to p), and hence is not open, by Theorem 3.9. QED

We are now prepared to state two very important theorems about open sets.

Theorem 3.11. The union of any collection of open sets is an open set.

Proof. Let H be an index set, and for every $p \in H$, let A_p be an open set. We want to show that $\bigcup_{p \in H} A_p$ is an open set. Thus let x be any point in $\bigcup_{p \in H} A_p$. By Definition 1.28, x must then be a member of at least one of the sets A_p, that is, there exists $p_o \in H$ such that $x \in A_{p_o}$. But A_{p_o} is an open set by hypothesis, so there exists an open interval (a, b) such that

$$x \in (a, b) \subset A_{p_o}$$

Since $A_{p_o} \subset \bigcup_{p \in H} A_p$ by Theorem 1.30, we thus have

$$x \in (a, b) \subset \bigcup_{p \in H} A_p \qquad \text{QED}$$

Theorem 3.12. The intersection of any finite collection of open sets is an open set.

Proof. Left as an exercise, using Theorem 3.8.

The next example shows that the word "finite" is essential in the statement of Theorem 3.12.

Example 3.13. Given any point p in R_1, define, for each $n \in I_o$ the open set $J_n = (p - 1/n, p + 1/n)$. It is left for the student to show that

$$\bigcap_{n \in I_o} J_n = \{p\}$$

But $\{p\}$ is not an open set, by Theorem 3.10. Therefore the intersection of an infinite collection of open sets need not be open.

Our next major theorem is to the effect that a subset H of R_1 is open iff H is the union of a countable collection of open intervals. To prove this theorem, we make use of the following theorem, and the notion of a component of an open set.

Theorem 3.14. Every open interval (a, b) in R_1 contains a rational number r.

Proof. We lose no generality if we assume that $a > 0$. Define $d = b - a$, so that $d > 0$. By the Archimedean property of R_1, there exists a positive integer N such that $Nd > 1$, that is, $1/N < d$. Also, by the Archimedean property, since $1/N > 0$, there exists a positive integer M such that $M(1/N) > b$; that is, $M/N > b$. Now, define H as the set of all those positive integers k such that $k/N \geq b$. Thus H is a nonempty set of positive integers since M is an element of H. By the Well-Ordering Axiom, H has a smallest element, which we denote by y. Thus $y/N \geq b$, and y is the smallest positive integer for which this inequality is true. It is easy to see that $y > 1$, since $1/N < d < b$ and $y(1/N) \geq b$.

Hence, for the positive integer $y - 1$, we must have $(y - 1)/N < b$. It remains to show that $(y - 1)/N > a$. Using various relations given, we see that

$$a = b - d \leq \frac{y}{N} - d < \frac{y}{N} - \frac{1}{N} = \frac{y - 1}{N}$$

Therefore the rational number $(y - 1)/N$ is between a and b.

QED

Theorem 3.14 tells us much more than might be apparent at first glance. Suppose we have two distinct real numbers a and b with $a < b$. The theorem then asserts the existence of a rational number r_1 such that $a < r_1 < b$. But now, a and r_1 are two distinct real numbers, so another application of the theorem yields a rational number r_2 such that $a < r_2 < r_1$. Similarly, r_1 and b are distinct real numbers; so again by Theorem 3.14, there exists a rational number r_3 such that $r_1 < r_3 < b$. We thus have three distinct rational numbers between a and b; that is, $a < r_2 < r_1 < r_3 < b$. This process can be continued indefinitely, yielding infinitely many distinct rational numbers between a and b. Therefore the assertion that we can find one rational number between a and b is equivalent to the statement that there are infinitely many distinct rational numbers between a and b. We now state this fact as a corollary to Theorem 3.14.

Corollary 3.15. Every open interval contains infinitely many distinct rational numbers.

Corollary 3.16. Every nonempty open subset H of R_1 contains infinitely many distinct rational numbers.

Definition 3.17. Let H be a subset of R_1, and p a point of H. The *component* C_p of H which contains p is the set consisting of the point p and all points q of H such that H contains every point between p and q.

Theorem 3.18. Let C_p be any component of a nonempty open subset H of R_1. Then C_p is a set of one of the following types:

(i) $C_p = R_1$, and here, $H = R_1$.
(ii) C_p is an open interval $(a, b) \subset H$, where $a \notin H$, $b \notin H$.
(iii) C_p is an open ray $\{x: x > a\} \subset H$, where $a \notin H$.
(iv) C_p is an open ray $\{x: a < x\} \subset H$, where $a \notin H$.

Proof. We prove this theorem in several steps. First, we show that C_p is a connected set. To do this, let a and b be any two points of C_p. By Definition 3.17, C_p contains the point a, the point b, the point p, every point between a and p, and every point between b and p. It is then a simple matter to prove that C_p contains every point between a and b, and hence is connected. We leave this proof as an exercise.

As the second step, we prove that C_p is open. Let q be any point in C_p. Then $q \in H$. Since H is open, there exists an open interval (a, b) such that $q \in (a, b) \subset H$. We leave it as an exercise to show that $(a, b) \subset C_p$. Therefore C_p is an open set.

As the third step, we show that if K is any connected subset of H such that $p \in K$, then $K \subset C_p$. To prove this, let x be any point of K. Since p and x are points of the connected subset K of H, every point between p and x is in H. Thus, by Definition 3.17, $x \in C_p$. Accordingly, $K \subset C_p$.

We have proven that C_p is a nonempty, open, connected subset of R_1. Consider the case where C_p is bounded, and let $a = \inf C_p$, $b = \sup C_p$. We must have $a < b$, by Theorem 3.10. Therefore, in our Catalogue of Connected Sets, C_p must be a set of type (3), (4), (5), or (6). Consequently the open interval $(a, b) \subset C_p$. Suppose that H contains both a and b. Then H contains the connected set $K = \{x : a \leq x \leq b\}$. By the third step of our proof, $K \subset C_p$. However, $C_p \subset K$, and hence $C_p = K$. This is impossible, since by Theorem 3.9, K is not an open set, but we know that C_p is an open set. This proves that H cannot contain both a and b. Similar arguments using sets of type (5) and (6) in our Catalogue of Connected Sets show that C_p cannot contain either a or b. Therefore, if C_p is bounded, then C_p is an open interval. Similar considerations of the cases where C_p is unbounded yield sets of type (i), (iii), or (iv). These are included in the exercises. QED

The result of our third step in the proof of Theorem 3.18 is sometimes reworded in the form: "A component of a nonempty open subset of R_1 is a maximal connected subset of H."

Theorem 3.19. If C_p is any component of an open subset H of R_1 and if x is any point of C_p, then $C_x = C_p$.

Proof. We prove that $C_x \subset C_p$, and leave the reverse inclusion to the student. Let $y \in C_x$. Then $y \in H$. Since $y \in C_x$, every point between x and y is in H. Since $x \in C_p$, every point between x and p is in H. Therefore, every point between y and p is in H. Since $y \in H$, we must have $y \in C_p$. QED

Corollary 3.20. If C_p and C_q are components of an open subset H of R_1 and if $C_p \cap C_q \neq \phi$, then $C_p = C_q$.

Proof. Let $z \in C_p \cap C_q$. By Theorem 3.19, $C_p = C_z = C_q$. QED

We have seen that if H is a nonempty open subset of R_1, then every component of H is a nonempty open set. Recalling the result of Corollary 3.16, we have at once the following lemma.

Lemma 3.21. If H is any nonempty open subset of R_1, then every component of H contains infinitely many distinct rational numbers.

We are now prepared to prove our principal theorem.

Theorem 3.22. A subset H of R_1 is open iff H is the union of a countable collection of open intervals.

Proof. If H is the union of a collection of open intervals, then H is open, by Theorem 3.11. Thus, we need only prove that given any nonempty open subset H of R_1, H is the union of a countable collection of open intervals. If we recall Theorem 2.7, Corollary 2.8, and Theorem 3.18, we see at once that every component of H is the union of a countable collection of open intervals. Thus the proof of our theorem will be complete if we can show that H has a countable number of components. We recall that the set of all rational numbers in R_1 is countable, and we denote by $\langle r_i \rangle$ the sequence consisting of all rational numbers contained in the set H. By Corollary 3.20, the rational number r_1 is contained in exactly one component of H. We denote this component by C_1. If $H = C_1$, H has a countable number of components, and our proof is complete. Suppose $H \neq C_1$. Then, by Lemma 3.21, there exists a smallest integer i_2 such that r_{i_2} is contained in a component of H distinct from C_1. Denote this component by C_2. Suppose we continue by this method and find k distinct components $C_1, C_2, \ldots, C_k$ of H. If $H = C_1 \cup C_2 \cup \ldots \cup C_k$, H is the union of a countable collection of components, and the proof is complete. If $H \neq C_1 \cup C_2 \cup \ldots \cup C_k$, then by Lemma 3.21, there exists a smallest positive integer i_{k+1} such that $r_{i_{k+1}}$ is not contained in any one of the components $C_1, C_2, \ldots, C_k$. Define C_{k+1} as the component of H containing $r_{i_{k+1}}$. If this process stops after a finite number of steps, then H is the union of a finite collection of components, and the proof is complete. Suppose that this is not so. Then we have defined inductively a sequence $\langle C_i \rangle$ of distinct components of H. It is to be noted that C_1 contains all rational numbers r_j such that $1 \leq j < i_2$, C_2 contains all rational numbers r_j such that $i_2 \leq j < i_3$, and, in general, C_k contains all rational numbers r_j such that $i_k \leq j < i_{k+1}$. Thus every rational number r_i in the sequence $\langle r_i \rangle$ is contained in exactly one of the components C_j which we have found in the sequence $\langle C_j \rangle$. If H has a component C which is not in this sequence, it follows at once that C contains no rational number of the sequence $\langle r_i \rangle$. This contradicts Lemma 3.21 and our definition of $\langle r_i \rangle$ as the sequence consisting of all rational numbers in the set H. QED

Before stating an important corollary of Theorem 3.22, we introduce the convenient terminology of disjoint sets.

Definition 3.23. Two sets H and K are said to be *disjoint* iff $H \cap K = \phi$. A collection of sets is said to be *disjoint* iff no two sets of the collection have a point in common.

Corollary 3.24. A subset H of R_1 is open iff H is the union of a countable collection of disjoint open connected sets.

One of the most important traits that an aspiring mathematician should try to develop is the ability to spot key theorems as he progresses through the development of a mathematical theory. Admittedly, every theorem worthy of the name has some importance. But certain theorems stand out as mile-markers of our progress. Some such theorems have been tagged "important" or "very important" in our preceding work. Our next theorem also rates special attention, although its importance may not be at all apparent to the reader. Before clarifying its importance, we may find it helpful to make another short digression into the realm of logic.

Suppose we have two given theorems A and B. Suppose further that we can prove that A implies B. By "A implies B" we mean "the truth of A implies the truth of B," or more simply, "if A is true, then B is true." Now suppose we can also prove that B implies A; in other words, the truth of either A or B implies the truth of the other. Then, we say that the theorems A and B are *logically equivalent*, or simply, *equivalent*. Thus, in particular, if we were to accept A as an axiom, we could prove B as a theorem; or if we were to accept B as an axiom, we could prove A as a theorem. In this sense, the next theorem is equivalent to the Least Upper Bound Axiom.

Theorem 3.25. There do not exist two disjoint nonempty open sets G_1 and G_2 such that $G_1 \cup G_2 = R_1$.

Proof. Suppose G_1 and G_2 are two such sets. Let C_p be a component of G_1. Since G_2 is not empty, $G_1 \neq R_1$; in particular, $C_p \neq R_1$. Therefore by Theorem 3.18, C_p must be a set of one of the following three types:

(1) C_p is an open interval $(a, b) \subset G_1$, where $a \notin G_1$, $b \notin G_1$.
(2) C_p is an open ray $\{x: x > a\} \subset G_1$, where $a \notin G_1$.
(3) C_p is an open ray $\{x: a < x\} \subset G_1$, where $a \notin G_1$.

Note that in all three of these cases, $a \notin G_1$; hence $a \in G_2$. This is impossible, since G_2 is an open set and it is clear that there is no open interval which contains a and is contained in G_2. QED

Although the proof of Theorem 3.25 as just given depends heavily on the use of the Least Upper Bound Axiom, in particular, on the existence of suprema and infima, it is also a fact that this theorem implies the Least Upper Bound Axiom. In other words, we could equally well have chosen the statement of Theorem 3.25 as an axiom, and then proven the Least Upper Bound Axiom as a theorem (this, of course, would have required a different arrangement of our work).

Theorem 3.26. If we accept as true the statement of Theorem 3.25, then every nonempty set H which is bounded above has a least upper bound.

Proof. Suppose that H has no least upper bound. Define the sets G_1 and G_2 as follows:

$$G_1 = \{x: x \text{ is an upper bound for } H\}$$
$$G_2 = \{x: x \text{ is not an upper bound for } H\}$$

Note that G_1 and G_2 are disjoint sets whose union is R_1. Our contrapositive proof will be complete if we can show that each of the sets G_1 and G_2 is nonempty and open. The set G_1 is nonempty since

H is bounded above. Let x be any element in G_1. Then x is not a least upper bound for H. There must thus exist a point $c < x$ such that c is an upper bound for H. It is clear that the open interval $(c, x + 1)$ contains x and is contained in G_1. Therefore G_1 is an open set. The proof that G_2 is a nonempty open set is left as an exercise. QED

PROBLEMS

18. Complete the proof of Theorem 3.8.
19. Prove Theorem 3.12.
20. Complete the discussion of Example 3.13.
21. Complete the proof of the first step of Theorem 3.18.
22. Complete the proof of the second step of Theorem 3.18.
23. Give the "similar arguments" mentioned near the end of Theorem 3.18.
24. Give the "similar considerations" mentioned near the end of Theorem 3.18.
25. Complete the proof of Theorem 3.19.
26. In the proof of Theorem 3.26, show that the set G_2 is both nonempty and open.
27. Let L_1 and L_2 be two disjoint nonempty sets whose union is R_1. Suppose that for any pair of points x and y such that $x \in L_1$, $y \in L_2$, we have $x < y$. Prove that either L_1 has a greatest element, or L_2 has a least element.
28. Define the following two subsets of R_1:

$$L_1 = \{x : x \geq 0, x^2 < 2\} \cup \{x : x < 0\}$$
$$L_2 = \{x : x > 0, x^2 > 2\}$$

Prove that each of the sets L_1 and L_2 is both nonempty and open. What conclusion can you draw from these facts? Do not use the square root of 2 anywhere in your proof.

4. *Closed sets. Compact sets*

We are now ready to define a closed set in R_1.

Definition 4.1. A set H in R_1 is a *closed set* iff its complement $C(H)$ is an open set.

Since the sets R_1 and ϕ are complementary sets (that is, each is the complement of the other), and since both ϕ and R_1 are open sets (by Theorem 3.3 and Corollary 3.7, respectively), we see as an immediate consequence of Definition 4.1 that each of these sets is also a closed set. These facts form part of the following stronger theorem.

Theorem 4.2. A subset H of R_1 is both open and closed iff $H = \phi$ or $H = R_1$.

Proof. We have seen that each of the sets ϕ and R_1 is both open and closed. Suppose that H is a subset of R_1 that is both open and closed. Then the set $C(H)$ is both open and closed, and $H \cup C(H) = R_1$. By Theorem 3.25, we must have either $H = \phi$, or $C(H) = \phi$, in which case $H = R_1$. QED

Theorem 4.3. The set $\{p\}$ is a closed set.

Proof. This is immediate, since the complement of $\{p\}$ is the union of two open rays, and hence is an open set. QED

In Theorem 1.16 of Chapter I, we proved DeMorgan's Laws for two sets. However, the theorem is true for an arbitrary collection of sets; we now state the theorem in its general form, and leave the proof as an exercise.

Theorem 4.4 (DeMorgan's Laws). Let U be the universe, H an index set, and $\{A_p\}$ a collection of subsets of U indexed by H. Then

(*a*) $$C\Big(\bigcup_{p \in H} A_p\Big) = \bigcap_{p \in H} C(A_p)$$

(*b*) $$C\Big(\bigcap_{p \in H} A_p\Big) = \bigcup_{p \in H} C(A_p)$$

Since closed sets are complements of open sets, many theorems about closed sets can be easily proved by working with the complementary open sets. The proofs of the following theorems can be effected by this technique, along with DeMorgan's Laws, and are left as exercises.

Theorem 4.5. The intersection of any collection of closed sets is a closed set.

Theorem 4.6. The union of any finite collection of closed sets is a closed set.

The student is requested in the exercises to prove by means of an example that the word "finite" is essential in Theorem 4.6. He should also show that sets of the type (4), (8), and (10) in the

Catalogue of Connected Sets are closed sets. Various types of sets in the catalogue are given special names.

Definition 4.7. A *closed interval* in R_1 is a set of the form $\{x: a \leq x \leq b\}$, where $a, b \in R_1$ with $a < b$. We normally denote the set as the closed interval $[a, b]$. The points a and b are called the *endpoints* of the closed interval $[a, b]$, and the positive number $b - a$ is called the *length* of the closed interval $[a, b]$.

Definition 4.8. A *half-open* (or *half-closed*) interval in R_1 is a set of the form $\{x: a < x \leq b\}$ or of the form $\{x: a \leq x < b\}$, where $a, b \in R_1$ with $a < b$. We normally denote these sets as the half-open (or half-closed) intervals $(a, b]$ and $[a, b)$, respectively. In either case, the points a and b are called the *endpoints*, and the positive number $b - a$ the *length*, of the half-open interval.

Closed rays may be similarly defined.

Definition 4.9. A *closed ray* in R_1 is a set of either the form $\{x: x \leq a\}$ or the form $\{x: a \leq x\}$, where $a \in R_1$. In either case, the point a is called the *endpoint* of the closed ray.

We thus have three types (open, half-open, and closed) of intervals in R_1, and two types (open and closed) of rays in R_1. All of these are connected sets in R_1, according to our Catalogue of Connected Sets. Moreover, except for the empty set ϕ, the set R_1 itself, and sets consisting of just one point, these various types of intervals and rays are the *only* connected sets in R_1.

We observe that each of the closed sets as just defined which is bounded below contains its infimum. This is a characteristic property of closed sets, which we establish as part of the following theorem.

Theorem 4.10. (*a*) Every closed set which is bounded below contains its infimum.

(*b*) Every closed set which is bounded above contains its supremum.

(*c*) Every closed and bounded set contains both its supremum and its infimum.

Proof. We prove (*a*), and leave (*b*) and (*c*) as exercises. Let H be any closed set which has a lower bound, and define $a = \inf H$. Suppose $a \notin H$. Then a is an element of the open set $C(H)$. Thus there exists an open interval (b, c) such that $a \in (b, c) \subset C(H)$.

But this means that c is a lower bound of H, which is impossible, since a is the greatest lower bound of H. QED

The next theorem is one of the most important in analysis. It is known as the Heine-Borel Theorem. We make use of one lemma in its proof.

Lemma 4.11. Let G be any open subset of R_1, and let p be any point of G. Then there exists an open interval (r, s) such that

$$p \in (r, s) \subset [r, s] \subset G$$

The proof is left as an exercise.

Theorem 4.12 (Heine-Borel Theorem). Let K be any closed and bounded subset of R_1, and let H be an index set. Suppose that for each $\alpha \in H$, there is an open set G_α and that $K \subset \bigcup_{\alpha \in H} G_\alpha$. Then there exists a finite subcollection $G_{\alpha_1}, G_{\alpha_2}, \ldots, G_{\alpha_k}$ of the collection $\{G_\alpha\}$ such that

$$K \subset G_{\alpha_1} \cup G_{\alpha_2} \cup \ldots \cup G_{\alpha_k}$$

Proof. If K is empty or a single point, then K is contained in one of the sets G_α, and the proof is complete. We may thus assume that K consists of at least two points. Define $a = \inf K$, and $b = \sup K$. By Theorem 4.10, $a \in K$ and $b \in K$, and since K contains at least two points, $a < b$. The proof now divides into two cases.

Case 1. K is the closed interval $[a, b]$.

Define the set L to consist of all points x of K satisfying the following condition:

(1) The closed interval $[a, x]$ is contained in the union of a finite number of the sets G_α.

Since $a \in K$, there exists an α_0 such that a is an element of the open set G_{α_0}. Thus there exists an open interval (y, x) such that $a \in (y, x) \subset G_{\alpha_0}$. Therefore, the closed interval $[a, x] \subset G_{\alpha_0}$. Consequently, the closed interval $[a, x] \subset L$; hence L is not empty.

The nonempty set L has b as an upper bound. By the Least Upper Bound Axiom, L has a supremum c. We note that $a < c \leq b$. We shall prove that $c = b$. Suppose that this is not so; that is, suppose that $a < c < b$. We know that there is an open set, which we denote by G_β, taken from the collection consisting of all the G_α, such that $c \in G_\beta$. By Lemma 4.11, there must exist an open interval (r, s) such that

$$c \in (r, s) \subset [r, s] \subset G_\beta$$

$a < r < s < b$. The point r is not an upper bound for L, since $r < c$. Therefore there exists a finite number of sets $G_{\beta_1}, G_{\beta_2}, \ldots, G_{\beta_n}$ of the collection consisting of all the G_α such that

$$[a, r] \subset G_{\beta_1} \cup G_{\beta_2} \cup \ldots \cup G_{\beta_n}$$

Since the closed interval $[r, s] \subset G_\beta$, and $c < s$, it follows that

$$[a, s] = [a, r] \cup [r, s] \subset G_\beta \cup G_{\beta_1} \cup G_{\beta_2} \cup \ldots \cup G_{\beta_n}$$

Consequently the point s is in the set L. This is impossible, since $c < s$, and $c = \sup L$. This contradiction proves that $c = b$. Therefore, there exists a finite number of sets $G_{\alpha_1}, G_{\alpha_2}, \ldots, G_{\alpha_k}$ of the collection $\{G_\alpha\}$ such that

$$K = [a, b] \subset G_{\alpha_1} \cup G_{\alpha_2} \cup \ldots \cup G_{\alpha_k}$$

This completes the proof of Case 1.

Case 2. K is not the closed interval $[a, b]$.

Define $G_o = C(K)$. Then the closed interval $[a, b]$ is contained in the union of the collection of open sets consisting of G_o and all of the sets G_α. Therefore, by Case 1, there is a finite subcollection $G_{\alpha_1}, G_{\alpha_2}, \ldots, G_{\alpha_k}$ of the G_α such that

$$K \subset [a, b] \subset G_o \cup G_{\alpha_1} \cup G_{\alpha_2} \cup \ldots \cup G_{\alpha_k}$$

Since G_o contains no point of K, we see that

$$K \subset G_{\alpha_1} \cup G_{\alpha_2} \cup \ldots \cup G_{\alpha_k} \qquad \text{QED}$$

The property of sets described in the Heine-Borel Theorem has acquired such importance in mathematics that it has been given a technical name. In order to introduce this terminology, we first define the notion of a covering of a set.

Definition 4.13. The collection of sets $\{G_\alpha\}$ is said to be a *covering* of the set K iff $K \subset \bigcup_{\alpha \in H} G_\alpha$, where H is the index set. This covering is said to be an *open covering* iff every G_α is an open set.

Definition 4.14. A set K is said to be *compact* iff every open covering of K contains a finite open subcovering of K. This means precisely that if $\{G_\alpha\}$ is a collection of open sets and if

$$K \subset \bigcup_{\alpha \in H} G_\alpha$$

where H is the index set, then there exists a finite subcollection $G_{\alpha_1}, G_{\alpha_2}, \ldots, G_{\alpha_k}$ of $\{G_\alpha\}$ such that

$$K \subset G_{\alpha_1} \cup G_{\alpha_2} \cup G_{\alpha_3} \cup \ldots \cup G_{\alpha_k}$$

We may now restate the Heine-Borel Theorem in the following form.

Theorem 4.15 (Heine-Borel Theorem). Every closed and bounded subset of R_1 is compact.

Our next two theorems will show that the converse of the Heine-Borel theorem is also true.

Theorem 4.16. Every compact set K is bounded.

Proof. For each $n \in I_o$, define the open interval $J_n = (-n, n)$. Note that $K \subset R_1 = \bigcup_{n \in I_o} J_n$. Therefore, since K is compact, there exists a positive integer N such that

$$K \subset J_1 \cup J_2 \cup \ldots \cup J_N = (-N, N)$$

It follows from this fact that for every x in K, we have $-N < x < N$. Accordingly, $-N$ is a lower bound for K, and N is an upper bound for K. Thus, K is bounded. QED

Theorem 4.17. Every compact set K is closed.

Proof. Let p be any point in $C(K)$. For each $n \in I_o$, define

$$H_n = \left[p - \frac{1}{n}, p + \frac{1}{n}\right]$$

It is left for the student to show that $\bigcap_{n \in I_o} H_n = \{p\}$. We observe that

$$K \subset C(\{p\}) = C(\bigcap_{n \in I_o} H_n) = \bigcup_{n \in I_o} C(H_n)$$

by DeMorgan's Theorem. Since K is compact and each of the sets $C(H_n)$ is open, there exists a positive integer N such that

$$\begin{aligned} K &\subset C(H_1) \cup C(H_2) \cup \ldots \cup C(H_N) \\ &= C(H_N) \end{aligned}$$

as the student can verify. Therefore $K \subset C([p - 1/N, p + 1/N])$, so that

$$p \in \left(p - \frac{1}{N}, p + \frac{1}{N}\right) \subset C(K)$$

We have proved that for any point p in $C(K)$, there exists an open interval containing p and contained in $C(K)$. Therefore $C(K)$ is an open set, hence K is closed. QED

We now combine the results of Theorems 4.15, 4.16, and 4.17 in the following summarizing theorem.

Theorem 4.18. A subset K of R_1 is compact iff K is both closed and bounded.

PROBLEMS

29. Prove Theorem 4.4.
30. Prove Theorem 4.5.
31. Prove Theorem 4.6.
32. Give an example of a collection of closed sets such that the union of these sets is not a closed set.
33. Show that sets of the type (4), (8), and (10) in the Catalogue of Connected Sets are closed.
34. Prove (*b*) and (*c*) of Theorem 4.10.
35. Prove Lemma 4.11.
36. Fill in the missing steps in the proof of Theorem 4.17.
37. Let K be a compact set, and H a closed set contained in K. Prove that H is compact.
38. Let $A = (0, 1]$, and define a collection of sets A_n as follows:

$$A_1 = (\tfrac{1}{3}, 2),\ A_2 = (\tfrac{1}{4}, \tfrac{1}{2}), \ldots,\ A_n = \left(\frac{1}{n+2}, \frac{1}{n}\right) \qquad \text{for } n \geq 2$$

(*a*) Prove that $\{A_n\}$ is an open covering of A.
(*b*) Prove that no finite subcollection of $\{A_n\}$ covers A. Thus conclude that A is not compact.
(*c*) Define $B = A \cup \{0\}$, and for each positive number b, define $B_b = (-1, b)$. Prove that the collection of sets consisting of all the A_n and B_b, for any fixed b, forms an open covering of B.
(*d*) Prove that, from the covering described in (*c*), we can find a finite number of sets which form an open covering of B.
(*e*) State why the result of (*d*) does not prove that B is a compact set.
39. Let $\{A_n\}$ be a collection of nonempty compact sets such that $A_n \supset A_{n+1}$ for every n. Prove that $\bigcap_{n \in I_0} A_n \neq \phi$. (*Hint:* Assume that $\bigcap_{n \in I_0} A_n = \phi$, and show that $\{C(A_n)\}$ is an open covering of A_1.)

IV
Convergence

1. Convergent sequences

In this section, we introduce and discuss the important concept of a convergent sequence of points. Our universe continues to be R_1, and the capital letters A, B, C, etc., which formerly designated only sets, are often used here as points in R_1. This ambiguity of notation is somewhat unfortunate, but it is the usual practice in mathematics since we have a limited number of letters at our disposal. It should cause no confusion, however, since the meaning of every letter introduced is either explicitly stated or is apparent from the context in which it is used.

Suppose we have a sequence $\langle a_n \rangle$ which is bounded and monotone nondecreasing. This means, of course, that

(i) $$a_1 \leq a_2 \leq a_3 \leq \ldots \leq a_n \leq \ldots \leq B$$

where B is an upper bound for the sequence. Since this sequence has an upper bound, it must have a supremum (by the Least Upper Bound Axiom), and we denote its supremum by A. It follows at once that $A \leq B$, and $a_n \leq A$ for every positive integer n. Thus, we can improve on (i) above by writing

(ii) $$a_1 \leq a_2 \leq a_3 \leq \ldots \leq a_n \leq \ldots \leq A$$

where A is the supremum of the sequence. Now, let ϵ be any positive number. Since ϵ is positive, we have

(iii) $$A - \epsilon < A < A + \epsilon$$

As an immediate consequence of (ii) and (iii), we see that

(iv) $a_n < A + \epsilon$ for every positive integer n

Now A is the supremum of the sequence, and by (iii), $A - \epsilon < A$. Therefore $A - \epsilon$ is not an upper bound for the sequence. This means that there must be at least one term of the sequence which is greater than $A - \epsilon$. If we denote such a term by a_{N+1}, we have

(v) $A - \epsilon < a_{N+1}$ for some positive integer N

But the sequence $\langle a_n \rangle$ is monotone nondecreasing; so

(vi) $a_{N+1} \leq a_n$ for every integer $n > N$

Combining (v) and (vi) yields

(vii) $A - \epsilon < a_n$ for every integer $n > N$

Finally, combining (iv) and (vii), we obtain

(viii) $A - \epsilon < a_n < A + \epsilon$ for every integer $n > N$

With the aid of Lemma 3.12 in Chapter I, (viii) can be written as

(ix) $|a_n - A| < \epsilon$ for every integer $n > N$

We have thus proved the following lemma.

Lemma 1.1. Let $\langle a_n \rangle$ be any bounded monotone nondecreasing sequence, and let A be the supremum of this sequence. Then given any positive number ϵ, there exists a positive integer N such that $|a_n - A| < \epsilon$ for every integer $n > N$.

Before proceeding further, let us take a more careful look at Lemma 1.1. Just what is meant by the assertion in (ix)? Remembering that absolute value can be interpreted geometrically as undirected distance, we see that (ix) merely says the following: Each term of the sequence beyond the term a_N (that is, each term $a_{N+1}, a_{N+2}, a_{N+3}, \ldots$) has the property that its undirected distance from the point A is less than ϵ. The conclusion of the lemma then asserts that no matter what positive number ϵ we are given, we can find a positive integer N (and hence the corresponding term a_N of the sequence) such that each term beyond a_N has this property. A loose translation into nonmathematical English might be as follows: "We can get the terms of the sequence as close as desired to the point A provided we go out far enough in the sequence." Of course, the precision required in mathematics does not allow for the use of expressions like "as close as desired" or "far enough." Thus we answer the question "How close?" with "A distance less than ϵ, where ϵ is any preassigned positive number"; and we answer the question "How far?" with "Beyond the term a_N, where N is a positive integer."

A few remarks about (viii) also prove helpful. We can think of $A - \epsilon$ and $A + \epsilon$ as the endpoints of an open interval. Then (viii) says the following: Each of the points $a_{N+1}, a_{N+2}, a_{N+3}, \ldots$ belongs to the open interval $(A - \epsilon, A + \epsilon)$. This means that there are at most N terms of the sequence (namely, the terms $a_1, a_2, a_3, \ldots, a_N$) that do not belong to the open interval. Since N is a positive integer, the set $\{a_1, a_2, \ldots, a_N\}$ is a finite set. Thus, the conclusion of Lemma 1.1 could be stated as follows: "Then given any positive number ϵ, the open interval $(A - \epsilon, A + \epsilon)$ contains all but a finite number of terms of the sequence."

Let us review what we have accomplished so far in this section. As a result of Lemma 1.1, we now know that every sequence $\langle a_n \rangle$ of real numbers which is bounded and monotone nondecreasing has associated with it a real number A (the supremum of the sequence) with the following property:

(1) For any $\epsilon > 0$, there exists a positive integer N such that

$$|a_n - A| < \epsilon \quad \text{for every integer } n > N$$

The order of choice of ϵ and N is of the greatest importance. We have referred to ϵ as a preassigned positive number. This means that $\epsilon > 0$ is to be chosen before N is determined. In general, the value of N depends on the choice of ϵ.

We now pose the following question: Are bounded, monotone nondecreasing sequences the *only* sequences of real numbers which enjoy property (1)? In other words, given an arbitrary (not necessarily bounded, and not necessarily monotone nondecreasing) sequence of real numbers, is it possible to find associated with this sequence a real number A (not necessarily the supremum of the sequence) such that property (1) is satisfied? The remainder of the section is devoted primarily to answering this question. Those sequences for which the answer is "yes" are of fundamental importance in analysis and are called convergent sequences.

Definition 1.2. The sequence $\langle a_n \rangle$ *converges* to the point A iff given any $\epsilon > 0$, there exists a positive integer N such that

$|a_n - A| < \epsilon$ for every integer $n > N$. A sequence is said to be *convergent* iff there exists a point A such that $\langle a_n \rangle$ converges to A. The point A is called the *limit* of the sequence $\langle a_n \rangle$, and we write $\lim a_n = A$.

We may now use this new terminology to restate Lemma 1.1 as a theorem in the following form.

Theorem 1.3. Every bounded monotone nondecreasing sequence of points in R_1 converges to its supremum.

The proof of the following theorem is analogous to the proof of Lemma 1.1, and is left as an exercise for the student.

Theorem 1.4. Every bounded monotone nonincreasing sequence of points in R_1 converges to its infimum.

Combining the results of Theorems 1.3 and 1.4, we have the following important theorem.

Theorem 1.5. In R_1, every bounded monotone sequence converges.

The remarks following Lemma 1.1 apply equally well to the statement of Definition 1.2. It is suggested that the student read these remarks again carefully, with particular attention to the restatement of the conclusion of Lemma 1.1. It should then be evident that Definition 1.2 could be similarly stated. Because of its importance in our future work, this equivalent form of Definition 1.2 is now listed as a theorem.

Theorem 1.6. The sequence $\langle a_n \rangle$ converges to the point A iff given any $\epsilon > 0$, the open interval $(A - \epsilon, A + \epsilon)$ contains all but a finite number of terms of the sequence $\langle a_n \rangle$.

We already know that the supremum and the infimum of a sequence are unique when they exist. Thus a bounded monotone sequence has a unique limit, by Theorems 1.3 and 1.4. We would certainly hope that *any* convergent sequence has a unique limit, since otherwise we could not talk about *the* limit of the sequence. A simple application of Theorem 1.6 easily establishes the uniqueness of limits.

Theorem 1.7. A convergent sequence has a unique limit.

Proof. Let $\langle a_n \rangle$ be a convergent sequence with $\lim a_n = A$, and let B be any real number distinct from A. Since $B \neq A$,

$|B - A| > 0$. Thus, let $r = |B - A|$, so that $r > 0$. We are given that $\lim a_n = A$, so by Theorem 1.6, the open interval $(A - r/2, A + r/2)$ must contain all but a finite number of terms of the sequence $\langle a_n \rangle$. We leave it as an exercise for the student to show that the open intervals $(A - r/2, A + r/2)$ and $(B - r/2, B + r/2)$ are disjoint; hence the open interval $(B - r/2, B + r/2)$ cannot possibly contain all but a finite number of terms of the sequence. Thus, by Theorem 1.6, the sequence $\langle a_n \rangle$ does not converge to B. QED

The following lemma is fundamental in dealing with subsequences. The easy proof is left as an exercise for the student.

Lemma 1.8. If $\langle n_i \rangle$ is a subsequence of I_o, then $n_i \geq i$ for every positive integer i.

Theorem 1.9. Every subsequence of a convergent sequence in R_1 is convergent, and converges to the same point as the given sequence.

Proof. Let $\langle a_n \rangle$ be a convergent sequence with $\lim a_n = A$, and let $\langle b_i \rangle$ be any subsequence of $\langle a_n \rangle$. We want to show that $\lim b_i = A$. Let $\epsilon > 0$ be given. Since $\lim a_n = A$, there exists a positive integer N such that $|a_n - A| < \epsilon$ for every integer $n > N$. Now we look at the subsequence $\langle b_i \rangle$. By definition of a subsequence, we know there exists a subsequence $\langle n_i \rangle$ of I_o such that $b_i = a_{n_i}$ for each positive integer i. But for every integer $i > N$, we have $n_i > N$ by Lemma 1.8. Hence for every integer $i > N$, we must have $|a_{n_i} - A| < \epsilon$. Using the fact that $b_i = a_{n_i}$, it follows that $|b_i - A| < \epsilon$ for every integer $i > N$. Therefore $\lim b_i = A$. QED

It should be carefully noted that in Definition 1.2 (and in similar definitions throughout mathematics where the phrase "for any $\epsilon > 0$" appears), we may pick ϵ as any positive number whatever. In particular, if ϵ is any positive number, then $\epsilon/2$ is also positive. Thus, we can reword Definition 1.2 as follows: "The sequence $\langle a_n \rangle$ converges to the point A iff given any $\epsilon > 0$, there exists a positive integer N such that $|a_n - A| < \epsilon/2$ for every integer $n > N$." Of course, we should realize that the positive integer N in this statement is not necessarily the same as the value of N in Definition 1.2. In precisely the same way, the $\epsilon/2$ in the preceding statement could be replaced by any other positive number provided that N were changed accordingly. With this background in mind, we are now prepared to prove the following important theorem.

Theorem 1.10. Let $\langle a_n \rangle$ be any convergent sequence in R_1, and let $\lim a_n = A$. Then given any $\epsilon > 0$, there exists a positive integer N such that for any pair of integers h, k with $h > N$ and $k > N$, we have $|a_h - a_k| < \epsilon$.

Proof. Let $\epsilon > 0$ be given. Since $\lim a_n = A$, there exists a positive integer N such that $|a_n - A| < \epsilon/2$ for every integer $n > N$. Then, in particular, we have $|a_h - A| < \epsilon/2$ for $h > N$, and also $|a_k - A| < \epsilon/2$ for $k > N$. Thus for any pair of integers h, k where both $h > N$ and $k > N$, we must have

$$|a_h - a_k| = |a_h - A + A - a_k| \leq |a_h - A| + |A - a_k| < \frac{\epsilon}{2} + \frac{\epsilon}{2} = \epsilon$$

Therefore $|a_h - a_k| < \epsilon$. QED

The statement of Theorem 1.10 should come as no surprise to us, provided our intuitive notion of a convergent sequence is taking proper shape. Roughly speaking, the theorem says that if a sequence converges, we can get any pair of its terms as close together as desired provided we go out far enough in the sequence. Again, for mathematical rigor, the phrases "as close together as desired" and "go out far enough" are stated precisely in the theorem in terms of ϵ and N, respectively. Even more interesting at the moment is the neat technique of proof, which practically suggests itself. Our basic line of reasoning is as follows: If we simultaneously squeeze two terms of the sequence close to the point A, we must at the same time be squeezing them close to each other. Note how this is done in the proof. Given $\epsilon > 0$, we first use the convergence of the sequence $\langle a_n \rangle$ to find a point in the sequence beyond which any two terms a_h and a_k must each be within the distance $\epsilon/2$ of A. We then show that these two terms must be within the distance ϵ of each other. To do this, we use a technique which might be called "adding zero and splitting by the triangle inequality." Of course, the zero which is added is expressed in the convenient form $(-A + A)$. This device is a basic technique used in many proofs of this type, and is used as needed in subsequent proofs without further comment.

Recalling the definition of a uniformly isolated sequence (Definition 2.1 in Chapter II), we immediately have the following corollary to Theorem 1.10.

Corollary 1.11. No convergent sequence in R_1 is uniformly isolated.

Theorem 1.12. No subsequence of a convergent sequence in R_1 is uniformly isolated.

Proof. Every subsequence of a convergent sequence in R_1 is itself a convergent sequence, by Theorem 1.9, and hence cannot be uniformly isolated, by Corollary 1.11. QED

Theorem 1.13. Every convergent sequence in R_1 is bounded.

Proof. A convergent sequence in R_1 cannot have a uniformly isolated subsequence, by Theorem 1.12, and hence must be bounded, by Problem 17 of Chapter II. QED

It should be evident that we have now settled the question of convergence as far as monotone sequences are concerned. By Theorem 1.5, every monotone sequence converges if it is bounded. By Theorem 1.13, every sequence (including, of course, every monotone sequence) which converges must be bounded. We thus have an "iff" condition for convergence of a monotone sequence, which we state as a theorem.

Theorem 1.14. A monotone sequence in R_1 converges iff it is bounded.

We must point out, however, that boundedness alone is not an "iff" condition for convergence of an arbitrary sequence in R_1. Although it is true that every convergent sequence is bounded, it is not true that every bounded sequence is convergent. The following example illustrates this fact.

Example 1.15. Let $\langle a_n \rangle$ be the sequence $\langle 1, 0, 1, 0, 1, 0, 1, \ldots \rangle$, which is defined formally as follows: For each positive integer n,

$$a_n = \begin{cases} 1 & \text{if } n \text{ is odd} \\ 0 & \text{if } n \text{ is even} \end{cases}$$

This sequence is bounded; in fact, $\inf a_n = 0$ and $\sup a_n = 1$. But the sequence is not convergent. Note that the subsequence $\langle b_i \rangle$ of $\langle a_n \rangle$ defined by $b_i = a_{2i}$ for each $i \in I_o$ converges to zero. The student should define a subsequence of $\langle a_n \rangle$ which converges to one. It follows from Theorem 1.9 that the sequence $\langle a_n \rangle$ does not converge.

Despite the fact that boundedness alone is not a sufficient condition to guarantee convergence of a sequence, it nevertheless is a strong enough property to yield the following result, which is one of the most important of analysis.

Theorem 1.16 (Bolzano-Weierstrass Theorem for Sequences). Every bounded sequence $\langle a_n \rangle$ in R_1 has a convergent subsequence.

Proof. By Theorem 1.10 of Chapter II, $\langle a_n \rangle$ has a subsequence $\langle b_i \rangle$ which is monotone. The sequence $\langle b_i \rangle$ is also bounded, by Theorem 1.3 of Chapter II. Therefore, $\langle b_i \rangle$ converges, by Theorem 1.5. QED

PROBLEMS

1. State precisely the negation of Definition 1.2.
2. Prove Theorem 1.4.
3. Prove Lemma 1.8.
4. Prove that the two open intervals defined in Theorem 1.7 are disjoint.
5. Given two disjoint open intervals and a sequence $\langle a_n \rangle$ converging to a point A in one of these intervals, prove that the other open interval can contain at most a finite number of terms of the sequence.
6. Let $\langle a_n \rangle$ be a convergent sequence, with $\lim a_n = A$, and let k be a real number. Define a new sequence $\langle b_n \rangle$ as follows: For each positive integer n, $b_n = a_n + k$. Then, prove that $\lim b_n = A + k$.
7. Let $\langle a_n \rangle$ be a convergent sequence, with $\lim a_n = A$. Then prove
(*a*) $\lim 3a_n = 3A$.
(*b*) $\lim (-1)a_n = -A$.
8. Given two sequences $\langle a_n \rangle$ and $\langle b_n \rangle$, and a real number $\epsilon > 0$. Suppose there exists N_1 such that if $n > N_1$, then $|a_n - A| < \epsilon/2$; and suppose there exists N_2 such that if $n > N_2$, then $|b_n - B| < \epsilon/2$. Define $N = \sup \{N_1, N_2\}$. Prove that if $n > N$, then

$$|(a_n + b_n) - (A + B)| = |(a_n - A) + (b_n - B)| < \epsilon$$

and that

$$|(a_n - b_n) - (A - B)| < \epsilon$$

9. Let k be a given constant, and let $\langle a_n \rangle$ be the sequence defined by $a_n = k$ for every positive integer n. Prove that $\lim a_n = k$.

10. Prove that if two subsequences of a given sequence converge to distinct limits, then the sequence does not converge.

11. Complete Example 1.15 by defining a subsequence of $\langle a_n \rangle$ which converges to one.

12. Prove that every finite set is bounded, and contains both its supremum and its infimum.

13. Prove that if A and B are nonempty bounded sets, $A \cup B$ is bounded. Prove also that

(*a*) $\sup (A \cup B) = \sup \{\sup A, \sup B\}$.

(*b*) $\inf (A \cup B) = \inf \{\inf A, \inf B\}$.

14. Give an alternate proof of Theorem 1.13, using Theorem 1.6 and the results of Problems 12 and 13.

15. Let A and B be two distinct real numbers, and $\langle b_n \rangle$ a sequence of real numbers such that $\lim b_n = B$. Suppose that for each $n \in I_o$, $b_n \neq A$. Prove that there exists an $\epsilon > 0$ such that the open interval $(A - \epsilon, A + \epsilon)$ contains no term of the sequence $\langle b_n \rangle$.

2. *Properties of limits*

The limit of a sequence in R_1 has been defined as the real number to which the sequence converges. It is well to remember then that the symbol "$\lim a_n$" is meaningful only when it is either expressly stated or understood that $\langle a_n \rangle$ is a convergent sequence. Our aim in this section is to investigate some of the important properties of limits. Two such properties have already been proved in Section 1. Theorem 1.7 tells us that the limit of a convergent sequence is unique, and Theorem 1.9 asserts that if the point A is the limit of a sequence $\langle a_n \rangle$, then A is also the limit of every subsequence of $\langle a_n \rangle$. The following results are easily proved, and are left as exercises.

Theorem 2.1. The alteration of a finite number of terms of a sequence does not affect convergence. In other words, if $\langle b_n \rangle$ is the sequence obtained by altering a finite number of terms of the sequence $\langle a_n \rangle$, then $\lim a_n = A$ iff $\lim b_n = A$.

Theorem 2.2. If, from some point on (that is, for all $n \geq N$, where N is a positive integer), every term a_n of the sequence $\langle a_n \rangle$ is equal to the number k, then $\lim a_n = k$.

Note that Problem 9 is merely the special case of Theorem 2.2 where $N = 1$.

Before stating the next theorem, we must define what we mean by a sum of two or more sequences.

Definition 2.3. The *sum of the sequences* $\langle a_n \rangle$ and $\langle b_n \rangle$ is the sequence $\langle k_n \rangle$ defined by $k_n = a_n + b_n$ for each $n \in I_o$, and we write $\langle k_n \rangle = \langle a_n + b_n \rangle$.

This definition extends by an induction process to the sum of any finite number of sequences. Suppose, for example, that we want the sum of the three sequences $\langle a_n \rangle$, $\langle b_n \rangle$, and $\langle c_n \rangle$. We first apply the definition to the two sequences $\langle a_n \rangle$ and $\langle b_n \rangle$ to obtain $\langle a_n + b_n \rangle$; then we apply it again to the two sequences $\langle a_n + b_n \rangle$ and $\langle c_n \rangle$, and obtain $\langle a_n + b_n + c_n \rangle$.

Theorem 2.4. If $\lim a_n = A$ and $\lim b_n = B$, then

$$\lim (a_n + b_n) = A + B$$

Proof. Let $\epsilon > 0$ be given. Then, by definition of a limit, we want to find a positive integer N such that $|(a_n + b_n) - (A + B)| < \epsilon$ for all $n > N$. Since $\lim a_n = A$, there exists N_1 such that $|a_n - A| < \epsilon/2$ for all $n > N_1$; and since $\lim b_n = B$, there exists N_2 such that $|b_n - B| < \epsilon/2$ for all $n > N_2$. Problem 8 then completes the proof. QED

Corollary 2.5. If $\lim a_n = A$, and if k is any constant, then

$$\lim (a_n + k) = A + k$$

Proof. The sequence $\langle a_n + k \rangle$ is the sum of the two sequences $\langle a_n \rangle$ and $\langle b_n \rangle$, where $b_n = k$ for every n. The former converges to A by hypothesis, and the latter converges to k by Theorem 2.2. Hence their sum converges to $A + k$, by Theorem 2.4. QED

Compare Corollary 2.5 with Problem 6.

Theorem 2.4 remains true for any finite number m of convergent sequences, and the case $m = 3$ is included among the exercises. Noting that the conclusion of Theorem 2.4 may be written as $\lim (a_n + b_n) = \lim a_n + \lim b_n$, we can state the theorem in words as follows: The sum of two convergent sequences is a convergent sequence, and the limit of the sum is equal to the sum of the limits.

We should also point out here that Theorem 2.4 is not an "iff" statement; that is, two sequences which do not converge may have a sum that does converge, as shown by the following example.

Example 2.6. Let $\langle a_n \rangle$ be the sequence $\langle 1, 0, 1, 0, 1, 0, \ldots \rangle$, and let $\langle b_n \rangle$ be the sequence $\langle 0, 1, 0, 1, 0, 1, \ldots \rangle$. Then neither sequence converges (see Example 1.15). But their sum is the sequence $\langle 1, 1, 1, 1, \ldots \rangle$, which does converge, by Theorem 2.2.

We are now ready to discuss products of sequences.

Definition 2.7. The *product of the sequences* $\langle a_n \rangle$ and $\langle b_n \rangle$ is the sequence $\langle k_n \rangle$ defined by $k_n = a_n b_n$ for each $n \in I_o$, and we write $\langle k_n \rangle = \langle a_n b_n \rangle$.

This definition can also be extended by an induction process to the product of any finite number of sequences.

Theorem 2.8. If $\lim a_n = A$ and $\lim b_n = B$, then

$$\lim a_n b_n = AB$$

Proof. Let $\epsilon > 0$ be given. Then we want to find N such that $|a_n b_n - AB| < \epsilon$ for every $n > N$. We employ the technique of "adding zero and splitting" to get

$$\begin{aligned} |a_n b_n - AB| &= |a_n b_n - a_n B + a_n B - AB| \\ &= |a_n(b_n - B) + B(a_n - A)| \\ &\leq |a_n(b_n - B)| + |B(a_n - A)| \\ &= |a_n| \cdot |b_n - B| + |B| \cdot |a_n - A| \end{aligned}$$

Now our problem is to get this last expression less than ϵ. We are given that $\langle a_n \rangle$ converges, and hence by Theorem 1.13, $\langle a_n \rangle$ is bounded. This means that there exists a real number $K > 0$ such that $|a_n| \leq K$ for every n, by Theorem 1.2 of Chapter II. Since K is positive, we see that $\epsilon/(K + |B|)$ is a positive number. We now use the fact that $\lim a_n = A$ and $\lim b_n = B$. There exists a positive integer N_1 such that $|a_n - A| < \epsilon/(K + |B|)$ for every $n > N_1$, and there exists a positive integer N_2 such that $|b_n - B| < \epsilon/(K + |B|)$ for every $n > N_2$. Define $N = \sup \{N_1, N_2\}$. Then for all $n > N$, we have

$$\begin{aligned} |a_n b_n - AB| &\leq |a_n| \cdot |b_n - B| + |B| \cdot |a_n - A| \\ &< K \cdot \frac{\epsilon}{K + |B|} + |B| \cdot \frac{\epsilon}{K + |B|} \\ &= (K + |B|) \cdot \frac{\epsilon}{K + |B|} \\ &= \epsilon \end{aligned}$$

QED

Theorem 2.8 remains true for any finite number m of convergent sequences, and the case $m = 3$ is included among the exercises.

Corollary 2.9. If $\lim a_n = A$, and if k is any constant, then

$$\lim ka_n = kA$$

Proof. The sequence $\langle ka_n \rangle$ is the product of the two sequences $\langle b_n \rangle$, where $b_n = k$ for each n, and $\langle a_n \rangle$. The former converges to k by Theorem 2.2, and the latter converges to A by hypothesis. Hence their product converges to kA, by Theorem 2.8. QED

Note that the limits determined in Problem 7 are merely special cases of Corollary 2.9.

Definition 2.10. The *difference of the sequences* $\langle a_n \rangle$ and $\langle b_n \rangle$ is the sequence $\langle k_n \rangle$ defined by $k_n = a_n - b_n$ for each $n \in I_o$, and we write $\langle k_n \rangle = \langle a_n - b_n \rangle$.

Theorem 2.11. If $\lim a_n = A$ and $\lim b_n = B$, then

$$\lim (a_n - b_n) = A - B$$

Proof. For each $n \in I_o$, $(a_n - b_n) = [a_n + (-1)b_n]$. Since $\lim (-1)b_n = -B$ by Corollary 2.9, we thus have, by Theorem 2.4,

$$\begin{aligned} \lim (a_n - b_n) &= \lim [a_n + (-1)b_n] \\ &= \lim a_n + \lim (-1)b_n \\ &= A + (-B) \\ &= A - B \end{aligned}$$

QED

Our next concern, naturally enough, will be quotients of sequences. Here, just as in algebra, a small complication arises, since division by zero is impossible.

Definition 2.12. Let $\langle b_n \rangle$ be a sequence, all of whose terms are nonzero. Then, the *quotient of the sequences* $\langle a_n \rangle$ and $\langle b_n \rangle$ is the sequence $\langle k_n \rangle$ defined by $k_n = a_n/b_n$ for each $n \in I_o$, and we write $\langle k_n \rangle = \langle a_n/b_n \rangle$.

Lemma 2.13. If $\lim b_n = B$, where each term $b_n \neq 0$ and $B \neq 0$, then the sequence $\langle 1/b_n \rangle$ is bounded.

Proof. By Problem 15, with $A = 0$, there exists an $\epsilon > 0$ such that the open interval $(-\epsilon, \epsilon)$ contains no term of the sequence $\langle b_n \rangle$. In terms of absolute values, this means that $|b_n| \geq \epsilon$ for every positive integer n. Therefore we have $|1/b_n| \leq 1/\epsilon$ for every n, hence $\langle 1/b_n \rangle$ is bounded, by Theorem 1.2 of Chapter II. QED

Theorem 2.14. If $\lim b_n = B$, where each term $b_n \neq 0$ and $B \neq 0$, then

$$\lim \frac{1}{b_n} = \frac{1}{B}$$

Proof. Let $\epsilon > 0$ be given. We want to find a positive integer N such that $|1/b_n - 1/B| < \epsilon$ for every $n > N$. Note by Lemma 2.13 that there exists a real number $K > 0$ such that $|1/b_n| \leq K$ for every n. Since $\lim b_n = B$, there exists a positive integer N such that $|B - b_n| < (|B|/K)\cdot\epsilon$. Hence, for all $n > N$, we have

$$\begin{aligned}\left|\frac{1}{b_n} - \frac{1}{B}\right| &= \left|\frac{B - b_n}{b_n B}\right| \\ &= \frac{|B - b_n|}{|B|}\cdot\frac{1}{|b_n|} \\ &= \frac{1}{|b_n|}\cdot\frac{1}{|B|}\cdot|B - b_n| \\ &< K\cdot\frac{1}{|B|}\cdot\frac{|B|}{K}\cdot\epsilon \\ &= \epsilon \end{aligned}$$

QED

Theorem 2.15. If $\lim a_n = A$ and $\lim b_n = B$, where each term $b_n \neq 0$ and $B \neq 0$, then

$$\lim \frac{a_n}{b_n} = \frac{A}{B}$$

Proof. For each n, we may write $a_n/b_n = a_n(1/b_n)$. Since $\lim 1/b_n = 1/B$ by Theorem 2.14, we thus have, by Theorem 2.8,

$$\lim \frac{a_n}{b_n} = \lim\left(a_n\cdot\frac{1}{b_n}\right) = (\lim a_n)\left(\lim\frac{1}{b_n}\right) = A\left(\frac{1}{B}\right) = \frac{A}{B}$$

QED

We thus see that convergent sequences and their limits behave quite properly when subjected to the fundamental operations of arithmetic. Many additional properties of limits are worthy of our attention; some of these are listed as theorems, whereas others are included among the exercises.

The following theorem is merely a restatement of the Archimedean property of R_1, and its proof is left as an exercise.

Theorem 2.16. $\text{Lim}\,\dfrac{1}{n} = 0$.

Two important properties of limits should be apparent from Theorem 2.16, although they could easily escape our attention. First, the limit (zero) of the sequence $\langle 1/n \rangle$ is not a member of the sequence, since for every positive integer n, $1/n > 0$. On the other hand, the limit k of the sequence $\langle a_n \rangle$, where $a_n = k$ for each n, in Theorem 2.2, is a member of the sequence. In general, then, *the limit of a sequence may, but need not, belong to the sequence.* Second, note that the sequences $\left\langle \frac{1}{n} \right\rangle$ and $\left\langle \frac{1}{2n} \right\rangle$ have the property that each term of the former sequence is greater than (in fact, twice as large as) the corresponding term of the latter sequence; yet they have the *same* limit. We now generalize this result in the following theorem.

Theorem 2.17. If $\lim a_n = A$, $\lim b_n = B$, and if $a_n > b_n$ for every $n \in I_o$, then $A \geq B$.

Proof. Left as an exercise. Note, however, that the critical point here is that we must allow for equality of the limits.

Theorem 2.18. If $0 \leq |r| < 1$, then $\lim r^n = 0$.

Proof. By Problem 30, we lose no generality if we assume that $0 < r < 1$. It follows at once that the sequence $\langle r^n \rangle$ is monotone decreasing and has zero as a lower bound. By Theorem 1.5, this sequence converges to a real number A. By Theorem 1.9, the subsequence $\langle r^{2n} \rangle$ of $\langle r^n \rangle$ must also converge to A. Accordingly, by Corollary 2.9, we must have

$$A = \lim r^{2n} = \lim (r^n \cdot r^n) = (\lim r^n)(\lim r^n) = A^2$$

Therefore, $A = A^2$, so that we must have $A = 0$ or $A = 1$. We cannot have $A = 1$ since the open interval $(r, 2)$ contains 1 but contains no term of the sequence $\langle r^n \rangle$. Therefore $A = 0$. QED

PROBLEMS

16. Prove Theorem 2.1.

17. Prove Theorem 2.2 in each of the following two ways:

(*a*) Use Theorem 2.1 and Problem 9.

(*b*) Use Theorem 1.6.

18. Prove the extension of Theorem 2.4 to three convergent sequences; that is, if $\lim a_n = A$, $\lim b_n = B$, and $\lim c_n = C$, then

$$\lim (a_n + b_n + c_n) = A + B + C$$

19. Prove the extension of Theorem 2.8 to three convergent sequences; that is, if $\lim a_n = A$, $\lim b_n = B$, and $\lim c_n = C$, then

$$\lim (a_n b_n c_n) = ABC$$

20. State Theorem 2.8 in words without using any symbols.
21. Show that Theorem 2.8 is not an "iff" statement by giving an example of two nonconvergent sequences whose product converges.
22. Prove: If $\lim a_n = 0$ and $\langle b_n \rangle$ is bounded, then $\lim (a_n b_n) = 0$.
23. State Theorem 2.11 in words without using any symbols.
24. Show that Theorem 2.11 is not an "iff" statement by giving an example of two nonconvergent sequences whose difference converges.
25. State Theorem 2.15 in words without using any symbols.
26. Show that Theorem 2.15 is not an "iff" statement by giving an example of two nonconvergent sequences whose quotient is a convergent sequence.
27. Prove Theorem 2.16.
28. Prove Theorem 2.17.
29. Prove that if $\langle a_n \rangle$ is a convergent sequence in R_1, then the three numbers $\inf a_n$, $\lim a_n$, and $\sup a_n$ all exist. Furthermore,

$$\inf a_n \leq \lim a_n \leq \sup a_n$$

30. Prove that $\lim a_n = 0$ iff $\lim |a_n| = 0$. (*Hint:* Use Problem 22.)
31. Prove that if $\lim a_n = A$, then $\lim |a_n| = |A|$. (*Hint:* Use (9), Theorem 3.13, Chapter I.) Is the converse true? If so, prove it; if not, give a counterexample.
32. Prove that the sequence $\left\langle \frac{1}{n} \right\rangle$ is monotone decreasing and bounded. Use the subsequence $\left\langle \frac{1}{2n} \right\rangle$ and a technique similar to that used in the proof of Theorem 2.18 to get a proof of Theorem 2.16 which does not make direct use of the Archimedean property.

3. Cauchy sequences

Suppose we are given a sequence $\langle a_n \rangle$ of real numbers, and we want to determine whether or not the sequence is convergent. In some cases, the answer is an immediate consequence of theorems which have already been proved. For example, if the sequence $\langle a_n \rangle$ is not bounded, then it cannot converge. On the other hand, if the sequence $\langle a_n \rangle$ is bounded and also monotone, then it does converge, and its limit A is either $\sup a_n$ or $\inf a_n$, depending on whether the sequence is monotone nondecreasing or monotone nonincreasing, respectively. However, what if the sequence $\langle a_n \rangle$ is bounded but not monotone? Here the situation becomes critical, for a check of our work thus far shows that we have proved no theorem that answers the question of convergence in this case. Thus, our only hope is to fall back on the definition of convergence; that is, we try to find a real number A associated with the sequence $\langle a_n \rangle$ in such a way that Definition 1.2 is satisfied. But note the difficulty implied by this procedure. In order to use Definition 1.2 to prove that a given sequence converges, we must *know in advance* (or at least suspect beyond a reasonable doubt) the value of its limit. This is indeed a handicap, and often an unreasonable one, for in many cases we are interested only in the existence of a limit, not its value. Thus we seek some other criterion for convergence; that is, we want to find an "iff" condition for convergence of a sequence, and, if possible, one that does not involve the value of the limit. With this objective in mind, we now define a Cauchy sequence.

Definition 3.1. A sequence $\langle a_n \rangle$ in R_1 is said to be a *Cauchy sequence* iff given any $\epsilon > 0$, there exists a positive integer N such that for any pair of integers h, k, with $h > N$ and $k > N$, we have $|a_h - a_k| < \epsilon$.

By virtue of this definition, we may now restate Theorem 1.10 as follows.

Theorem 3.2. Every convergent sequence in R_1 is a Cauchy sequence.

The proof of the following theorem is left as an exercise.

Theorem 3.3. Every Cauchy sequence in R_1 is bounded.

With the aid of Theorem 1.16, we now get the following corollary to Theorem 3.3.

Corollary 3.4. Every Cauchy sequence in R_1 has a convergent subsequence.

Theorem 3.5. Let $\langle a_n \rangle$ be a Cauchy sequence in R_1, and let $\langle b_i \rangle$ be a subsequence of $\langle a_n \rangle$. If $\langle b_i \rangle$ is convergent and $\lim b_i = B$, then $\langle a_n \rangle$ is convergent and $\lim a_n = B$.

Proof. Let $\epsilon > 0$ be given. Since $\langle a_n \rangle$ is a Cauchy sequence, we know that

(1) There exists a positive integer N_1 such that if $h > N_1$ and $k > N_1$, $|a_h - a_k| < \epsilon/2$.

Also, since $\lim b_i = B$ by hypothesis, we know that

(2) There exists a positive integer N_2 such that for all integers $i > N_2$, $|B - b_i| < \epsilon/2$.

Now define $N = N_1 + N_2$. Then $N > N_2$, so by (2) we have $|B - b_N| < \epsilon/2$. Also, since $\langle b_i \rangle$ is a subsequence of $\langle a_n \rangle$, there exists a positive integer M such that $a_M = b_N$. Furthermore, $M \geq N$ by Lemma 1.8, so $M > N_1$. Hence by (1), for any integer $n > N_1$, we have $|a_M - a_n| < \epsilon/2$. Finally, for any integer $n > N$, it follows that

$$\begin{aligned} |B - a_n| &= |B - b_N + b_N - a_n| \\ &\leq |B - b_N| + |b_N - a_n| \\ &= |B - b_N| + |a_M - a_n| \\ &< \frac{\epsilon}{2} + \frac{\epsilon}{2} \\ &= \epsilon \end{aligned}$$

Therefore $\langle a_n \rangle$ is convergent, and $\lim a_n = B$. QED

Among other things, Theorem 3.5 asserts that if a Cauchy sequence has a convergent subsequence, the Cauchy sequence converges. Corollary 3.4, however, has already shown us that every Cauchy sequence has a convergent subsequence. We thus get the following very important result.

Theorem 3.6. Every Cauchy sequence in R_1 is convergent.

Combining Theorems 3.2 and 3.6, we see that we now have an "iff" condition for convergence of a sequence in R_1; furthermore, this condition is stated in terms of a Cauchy sequence, which does not involve the value of the limit.

Theorem 3.7 (Cauchy Criterion). A sequence in R_1 is convergent iff it is a Cauchy sequence.

We sometimes use the expression "R_1 is *complete*" to mean that every Cauchy sequence in R_1 converges. Theorem 3.6 thus asserts that R_1 is complete. It is worth noting that we could have accepted this fact as an axiom, known as the Axiom of Completeness, and then used this axiom to prove the Least Upper Bound Property. In other words, we now have another property, that of completeness of R_1, which is equivalent to the Least Upper Bound Axiom. This equivalence is mentioned here only as a matter of interest, and is not proved.

PROBLEMS

33. State precisely what is meant by saying that a sequence $\langle a_n \rangle$ in R_1 is not a Cauchy sequence.
34. Prove that every subsequence of a Cauchy sequence is a Cauchy sequence.
35. Prove that no Cauchy sequence in R_1 is uniformly isolated, and no subsequence of a Cauchy sequence in R_1 is uniformly isolated.
36. Prove Theorem 3.3.
37. Prove: the real number b is the supremum of the sequence $\langle x_n \rangle$ in R_1 iff given any $\epsilon > 0$, the following conditions are satisfied:
(*a*) $x_n < b + \epsilon$ for every n.
(*b*) $x_n > b - \epsilon$ for at least one n.
38. State and prove a corresponding theorem for infima.
39. Let H be a nonempty set having b as its supremum and suppose $b \notin H$. Prove that there exists a monotone nondecreasing sequence of distinct points of H which converges to b.
40. State and prove a corresponding result for infima.

4. *Neighborhoods, cluster points, and sequential compactness*

We shall find it convenient to introduce another name for nonempty open subsets of R_1. Any such set is called a neighborhood of each of its points.

Definition 4.1. By a *neighborhood* of a point p in R_1 we mean any open set containing p.

We recall from a theorem in Chapter III that a set H in R_1 is not open iff there exists a point p in H such that every open interval containing p contains at least one point not in H. This criterion for nonopen sets can be restated as in the following theorem, the proof of which is left as an exercise.

Theorem 4.2. A subset H of R_1 is not open iff there exists a point p in H such that every neighborhood of p contains at least one point not in H.

The theorem just stated motivates our definition for cluster points.

Definition 4.3. A point p is said to be a *cluster point* (or an *accumulation point*) of a subset H of R_1 iff every neighborhood of p contains at least one point of H distinct from p. If H is a subset of R_1, then by H' we denote the set of all cluster points of H. The set H' is called the *derived set* of H.

Example 4.4. Let H be the half-open interval

$$(a, b] = \{x: a < x \leq b\}$$

We shall prove that $H' = [a, b] = \{x: a \leq x \leq b\}$. To prove this, let p be any point in H, and let G be any neighborhood of p. Then there is an open interval (c, d) such that $p \in (c, d) \subset G$. Define $r = \sup \{a, c\}$, and note that $a \leq r < p \leq b$. The point $(p + r)/2$ is the midpoint of the line segment joining r and p. It must therefore be a point of the open interval (c, d). Therefore G contains the point $(p + r)/2$ of H, and this point is not the point p. Thus p is a cluster point of H, that is, $p \in H'$. Since p was an arbitrary point of H, we have proven that $H \subset H'$. We leave as an exercise for the student the proof that $a \in H'$. Thus $H' \supset [a, b]$. However, H cannot contain any point of the complement of $[a, b]$. To see this, suppose that $p \notin [a, b]$. If $p < a$, then the open interval $(p - 1, a)$ is a neighborhood of p containing no point of H. Similarly, if $p > b$, then the open interval $(b, p + 1)$ is a neighborhood of p containing no point of H. In either case, $p \notin H'$. Therefore $H' = [a, b]$.

We note in the example just given that the point b is a cluster point of H which is in H, whereas the point a is a cluster point of H which is not in H. This shows us that the statement "p is a cluster point of H" gives us no information whatever as to whether or not p is in H.

The discussion in Example 4.4 would have been slightly less involved if we had had the following theorem at our disposal. The proof is left as an exercise.

Theorem 4.5. A point p is a cluster point of a set H iff every open interval containing p contains at least one point of H distinct from p.

Our next theorem ties together the notions of cluster points and convergent sequences.

Theorem 4.6. A point p is a cluster point of a set H iff there exists a sequence of distinct points of H converging to p.

Proof. We must prove two things:

(1) If there is a sequence of distinct points of H converging to the point p, then p is a cluster point of H.

(2) If p is a cluster point of H, then there exists a sequence of distinct points of H converging to p.

The proof of (1) is left as an exercise. To prove (2), let p be any cluster point of H. Define the sets K and L as follows:

$$K = \{x: x \in H, x < p\}$$
$$L = \{x: x \in H, x > p\}$$

We shall prove that p must be either the supremum of K or the infimum of L. Suppose that this is not so. Then there is a number $c < p$ such that c is an upper bound for K, and there is a number $d > p$ which is a lower bound for L. It follows at once that the open interval (c, d) contains no point of H distinct from p. Therefore p is not a cluster point of H. This contradiction shows that we must have either $c = \sup K$ or $c = \inf L$. It follows from Problems 39 and 40 that in either case there is a monotone sequence of distinct points of H which converges to p.

Corollary 4.7. If $H' \neq \phi$, then H is an infinite set.

Corollary 4.8. No finite set has a cluster point.

A restatement of Theorem 4.2 in terms of cluster points provides us with an extremely useful characterization of closed sets.

Theorem 4.9. Let H be a subset of R_1. Then $C(H)$ is not open (that is, H is not closed) iff at least one cluster point of H is in $C(H)$.

Corollary 4.10. A subset H of R_1 is not closed iff at least one cluster point of H is not in H.

Corollary 4.11. A subset H of R_1 is closed iff every cluster point of H is in H.

Corollary 4.12. A subset H of R_1 is closed iff $H' \subset H$.

Corollary 4.13. A subset H of R_1 is closed iff $H = H \cup H'$.

Regardless of whether or not the set H is closed, we give a special name to the set $H \cup H'$.

Definition 4.14. If H is a subset of R_1, we define $\overline{H} = H \cup H'$. The set $\overline{H}$ is called the *closure* of the set H.

The proof of the following theorem is left as an exercise.

Theorem 4.15. If H is a subset of R_1, then each of the sets H' and $\overline{H}$ is a closed set.

Corollary 4.13 can be restated in the following convenient form.

Theorem 4.16. A subset H of R_1 is closed iff $H = \overline{H}$.

Some of the properties of closure are stated in the next theorem; a few additional properties are included among the exercises.

Theorem 4.17. Let A and B be sets in R_1. Then

(1) $\overline{\phi} = \phi$
(2) $\overline{\overline{A}} = \overline{A}$
(3) $A \subset \overline{A}$
(4) If $A \subset B$, then $\overline{A} \subset \overline{B}$
(5) $\overline{A \cup B} = \overline{A} \cup \overline{B}$
(6) $\overline{A \cap B} \subset \overline{A} \cap \overline{B}$

Proof. Since ϕ has no limit points, $\phi' = \phi$, so

$$\overline{\phi} = \phi \cup \phi' = \phi \cup \phi = \phi$$

which proves (1). For (2), we see that $\overline{A}$ is closed, by Theorem 4.15, and hence is equal to its closure, by Theorem 4.16. (3) is an immediate consequence of the definition of $\overline{A}$. The proofs of (4) and (5) follow from Problems 46 and 47. To prove (6), note that $A \cap B \subset A$ and $A \cap B \subset B$; hence by (4), $\overline{A \cap B} \subset \overline{A}$ and $\overline{A \cap B} \subset \overline{B}$, so that $\overline{A \cap B} \subset \overline{A} \cap \overline{B}$. QED

The following theorem is an interesting characterization of the closure of a set. Its proof is quite easy and is left as an exercise.

Theorem 4.18. The closure of a set A in R_1 is the smallest closed set containing A, in the sense that if F is any closed set containing A, then $\overline{A} \subset F$. In particular, the set $\overline{A}$ is the intersection of all closed sets containing A.

We have proven that a subset K of R_1 is closed and bounded iff it is compact. We now introduce another property of sets in R_1 which is equivalent to the property of compactness.

Definition 4.19. A subset H of R_1 is *sequentially compact* iff given any sequence $\langle p_n \rangle$ of points of H, there exists a subsequence $\langle p_{n_i} \rangle$ of $\langle p_n \rangle$ which converges to a point p of H.

We have proven in Theorem 2.6 of Chapter II that a set of real numbers is unbounded iff it contains a monotone sequence of points which is uniformly isolated. We have also seen that every subsequence of a uniformly isolated sequence is uniformly isolated. In particular, this means, by Corollary 1.11, that a uniformly isolated sequence of points can contain no convergent subsequence. It follows at once from this discussion that if a subset H of R_1 is unbounded, then it is not sequentially compact. We state the contrapositive of this result as our next theorem.

Theorem 4.20. Every sequentially compact set is bounded.

It is not true that every bounded set is sequentially compact.

Example 4.21. The open interval $H = (2, 4)$ is bounded, but not sequentially compact. To see this, consider the sequence $\langle a_n \rangle$ of points of H defined by $a_n = 2 + 1/n$ for each $n \in I_o$. Note that every subsequence of $\langle a_n \rangle$ converges to the point 2, which is not in H. Therefore no subsequence of $\langle a_n \rangle$ converges to a point of H. Thus H is not sequentially compact.

Theorem 4.22. Every sequentially compact set H in R_1 is closed.

Proof. Suppose H is not closed. Then there exists a cluster point p of H such that $p \notin H$. By Theorem 4.6, there exists a sequence $\langle p_n \rangle$ of distinct points of H which converges to p. Thus no subsequence of $\langle p_n \rangle$ can converge to a point of H. Therefore H is not sequentially compact. QED

The student can easily verify the fact that R_1 is a closed set which is not sequentially compact. From this result and Example 4.21, we see that neither of the statements "H is closed," "H is bounded," implies that H is sequentially compact. The next theorem shows that the combination of both these statements is equivalent to sequential compactness.

Theorem 4.23 (Bolzano-Weierstrass Theorem). A set H in R_1 is sequentially compact iff it is closed and bounded.

Proof. We see from Theorems 4.22 and 4.20 that every sequentially compact set is both closed and bounded. We must prove that if H is both closed and bounded, H is sequentially compact. It follows at once from Definition 4.19 that the empty set is sequentially compact, so we may assume $H \neq \phi$. Let $\langle a_n \rangle$ be any sequence of points of H. Then by the Bolzano-Weierstrass Theorem for sequences (Theorem 1.16), $\langle a_n \rangle$ has a subsequence $\langle b_i \rangle$, where $b_i = a_{n_i}$ for each $i \in I_o$, which converges to a point p of R_1. If $\langle b_i \rangle$ contains a subsequence $\langle b_{i_j} \rangle$ of distinct terms, then $\langle b_{i_j} \rangle$ converges to p, so that p is a cluster point of H, by Theorem 4.6. Therefore $p \in H$, since H is closed. If $\langle b_i \rangle$ does not contain a subsequence of distinct terms, $\langle b_i \rangle$ contains a subsequence $\langle b_{i_j} \rangle$ such that $b_{i_j} = p$ for every $j \in I_o$. The student should prove this. Consequently, in either case, $p \in H$. Therefore, H is sequentially compact. QED

Corollary 4.24. A subset H of R_1 is compact iff it is sequentially compact.

PROBLEMS

41. Prove Theorem 4.2.
42. In Example 4.4, prove that $a \in H'$.
43. Prove Theorem 4.5.
44. Prove part (1) of Theorem 4.6.
45. Prove Theorem 4.15.
46. Prove that if $A \subset B$, then $A' \subset B'$.
47. Prove that $(A \cap B)' = A' \cap B'$.
48. Prove Theorem 4.18.
49. Fill in the missing steps in the proof of Theorem 4.23.
50. Prove that if $\langle p_n \rangle$ is a sequence of points in $A \cap B$, then $\langle p_n \rangle$ has a subsequence in A, or $\langle p_n \rangle$ has a subsequence in B.
51. Use the result of Problem 50 to prove that any cluster point of the union of two sets is a cluster point of at least one of the sets. Deduce as a corollary that the union of two closed sets is closed.
52. Reprove the first statement of Problem 51 without using sequences. (*Hint:* Use the contrapositive technique.)
53. Let A be a connected set in R_1, and A^* any set such that $A \subset A^* \subset \overline{A}$. Prove that A^* is connected. This says in particular that the closure of a connected set is connected.
54. Let $\langle r_i \rangle$ be the sequence of all rational numbers. Given any real number p, prove that there exists a subsequence of $\langle r_i \rangle$ which converges to p.

55. Let H be any subset of R_1 such that $H' \neq \phi$. Prove that H contains a sequence of distinct points which is a Cauchy sequence.

56. A subset H of R_1 is sequentially compact iff given any infinite subset K of H, $K' \cap H \neq \phi$.

57. Prove that a subset H of R_1 is not sequentially compact iff there exists at least one sequence of distinct points of H which has no cluster point in H.

58. Let H be a nonempty set having B as its supremum, and suppose $B \notin H$. Prove that B is a cluster point of H.

59. State and prove a corresponding result for infima.

V

Continuity and uniform continuity

1. Continuous functions

We have already defined and discussed the notion of a mapping f from a set A into (or onto) a set B, where A and B are sets in a given universe. Now, in the special case (as in most of our work thus far) where the universe is R_1, we see that both A and B are subsets of R_1; that is, both the domain and the range of f are sets of real numbers. Since the image of x under f is a real number, we often speak of f as a *real valued function*, or more generally, a *real-valued mapping*. We often shorten this by referring to f as a *function*, although the terms "mapping" and "function" are used interchangeably.

We stated in Chapter III that one of the important objectives in analysis is the determination of those properties of sets which are preserved under certain types of mappings. As a case in point, we showed that onto mappings preserve countability. Since Chapter IV is devoted to a detailed study of convergent sequences, we might well wonder what kind of mapping preserves convergent sequences. We call such a mapping a continuous mapping and define it as follows.

Definition 1.1. Let $A \subset R_1$, let f be a function $f: A \to R_1$, and let p be any point in A. We say that f is *continuous at the point p* iff given any sequence of points $\langle p_n \rangle$ of A which converges to the point p, then the sequence of images $\langle f(p_n) \rangle$ converges to the point

$f(p)$. We say that f is *continuous on the set* A iff f is continuous at the point p for every p in A. The statement "f is continuous," where $f: A \to R_1$, means "f is continuous on A."

As an immediate consequence of Definition 1.1, we have the following two theorems.

Theorem 1.2. Every constant function is continuous on R_1.

Proof. Let $f: R_1 \to R_1$ be defined by $f(x) = k$ for every $x \in R_1$, where k is a real number. We want to show that f is continuous on R_1. Thus, let p be any point in R_1, and let $\langle p_n \rangle$ be any sequence of points in R_1 such that $\lim p_n = p$. By definition of f, we see that $f(p_n) = k$ for every n, and also, $f(p) = k$. Thus, we have $\lim f(p_n) = k = f(p)$. Therefore f is continuous at the point p. Since p was chosen as an arbitrary point in R_1, f is continuous on R_1. QED

Theorem 1.3. The identity mapping on R_1 is continuous.

Proof. Recall that the identity mapping $f: R_1 \to R_1$ is defined by $f(x) = x$ for all x in R_1. To prove that this mapping is continuous, let p be any point in R_1, and let $\langle p_n \rangle$ be any sequence of points in R_1 such that $\lim p_n = p$. Then since for each n, $f(p_n) = p_n$, and also $f(p) = p$, we have

$$\lim f(p_n) = \lim p_n = p = f(p) \qquad \text{QED}$$

Before stating the next theorem, we must define what is meant by the sum of two functions.

Definition 1.4. Let $A \subset R_1$, and let f and g be functions such that $f: A \to R_1$ and $g: A \to R_1$. Then the *sum of the functions* f and g is the function $h: A \to R_1$ defined by $h(x) = f(x) + g(x)$ for every $x \in A$.

This definition extends by an induction process to the sum of any finite number of functions.

Theorem 1.5. The sum of two continuous functions is continuous.

Proof. Let $A \subset R_1$, let f and g be continuous functions from A into R_1, and let h be the sum of these two functions. Let p be any point in A, and let $\langle p_n \rangle$ be any sequence of points in A such that $\lim p_n = p$. Since f is continuous at p, we have $\lim f(p_n) = f(p)$; and since g is continuous at p, we have $\lim g(p_n) = g(p)$. By definition of h, we see that $h(p_n) = f(p_n) + g(p_n)$ for each n. Therefore, by Theorem 2.4 of Chapter IV, we have

$$\begin{aligned}\lim h(p_n) &= \lim\, [f(p_n) + g(p_n)] \\ &= \lim f(p_n) + \lim g(p_n) \\ &= f(p) + g(p) = h(p)\end{aligned}$$

QED

Corollary 1.6. The sum of any finite number of continuous functions is continuous.

Definition 1.7. Let $A \subset R_1$, and let f and g be functions such that $f: A \to R_1$ and $g: A \to R_1$. Then, the *product of the functions* f and g is the function h defined by $h(x) = f(x) \cdot g(x)$ for every $x \in A$.

This definition also extends by an induction process to the product of any finite number of functions.

The proof of the following theorem is similar to the proof of Theorem 1.5 (except that it depends on Theorem 2.8 in Chapter IV), and is left as an exercise.

Theorem 1.8. The product of any finite number of continuous functions is continuous.

By virtue of Theorem 1.2, we have the following corollary to Theorem 1.8.

Corollary 1.9. Let $A \subset R_1$, let $f: A \to R_1$ be continuous on A, and let k be a real number. Then the mapping $h: A \to R_1$ defined by $h(x) = k \cdot f(x)$ for every x in A is continuous.

Definition 1.10. Let $A \subset R_1$, and let f and g be functions such that $f: A \to R_1$ and $g: A \to R_1$. Then, the *difference of the functions* f and g is the function h defined by $h(x) = f(x) - g(x)$ for every $x \in A$.

Theorem 1.11. The difference of two continuous functions is continuous.

Proof. Left as an exercise. Note that

$$f(x) - g(x) = f(x) + (-1) \cdot g(x)$$

for every x in A.

Definition 1.12. Let $A \subset R_1$, and let f and g be functions such that $f: A \to R_1$ and $g: A \to R_1$. Furthermore, suppose $g(x) \neq 0$ for

$x \in A$. Then the *quotient of the functions* f and g is the function h defined by $h(x) = f(x)/g(x)$ for every $x \in A$.

Theorem 1.13. The quotient of two continuous functions is continuous, provided the function in the denominator is never zero.

Proof. Left as an exercise.

We now determine a large class of continuous functions. We begin by defining a polynomial.

Definition 1.14. A *polynomial in* x is a function $f: R_1 \to R_1$ defined by

$$f(x) = a_0x^n + a_1x^{n-1} + a_2x^{n-2} + \dots + a_n$$

for every $x \in R_1$, where n is a nonnegative integer, and a_0, a_1, a_2, . . . , a_n are constants in R_1, with $a_0 \neq 0$. The integer n is called the *degree* of the polynomial, and for each integer i, $0 \leq i \leq n$, a_ix^{n-i} is called the *term of degree* $n - i$ of the polynomial, and a_i is referred to as the *coefficient* of this term.

As a result of Definition 1.14, we may thus consider a constant function as a polynomial of zero degree. By virtue of Theorem 1.2, we see that every polynomial of zero degree is continuous. We now proceed to show that every polynomial (of any degree) is continuous.

Lemma 1.15. Let $f: R_1 \to R_1$ be defined by $f(x) = ax^n$ for every $x \in R_1$, where n is a nonnegative integer and a is a real number. Then, f is continuous on R_1.

Proof. If $n = 0$ or $a = 0$, then f is a constant function, and hence is continuous, by Theorem 1.2. If $n \geq 1$, define the function $f_1: R_1 \to R_1$ by $f_1(x) = x$ for all $x \in R_1$, and let f_2 be the product of n factors, each equal to f_1. The function f_1 is continuous, by Theorem 1.3, and hence f_2 is continuous, by Theorem 1.8. Therefore f is continuous, by Corollary 1.9. QED

Theorem 1.16. Every polynomial is continuous.

Proof. A polynomial is, by definition, a finite sum of functions of the form ax^n, where a is a real number and n is a nonnegative integer. But each of these functions is continuous, by Lemma 1.15, and hence their sum is continuous, by Corollary 1.6. QED

While on the subject of polynomials, let us digress for a moment to obtain some interesting results. These will lead eventually to an alternate proof that all polynomials are continuous, working merely from the definition of continuity. Recall that in algebra we learned two important theorems concerning polynomials, namely, the Factor Theorem and the Remainder Theorem. These are listed as Theorem 1.18 and Corollary 1.19, respectively. To prove them, we will use the following lemma, the proof of which is left as an exercise.

Lemma 1.17. If n is a positive integer and p is any real number, then $x^n - p^n$ has $x - p$ as a factor.

Theorem 1.18. If f is a polynomial and p is any real number, then $f(x) - f(p)$ has $x - p$ as a factor; that is, $f(x) - f(p) = (x - p)Q(x)$, where Q is a polynomial. Thus $x - p$ is a factor of $f(x)$ iff $f(p) = 0$.

Proof. Let f be any polynomial. Then we may write

$$f(x) = a_0x^n + a_1x^{n-1} + a_2x^{n-2} + \ldots + a_{n-1}x + a_n$$

Hence,

$$f(p) = a_0p^n + a_1p^{n-1} + a_2p^{n-2} + \ldots + a_{n-1}p + a_n$$

so that

$$f(x) - f(p) = a_0(x^n - p^n) + a_1(x^{n-1} - p^{n-1}) + \ldots + a_{n-1}(x - p)$$

By Lemma 1.17, $x - p$ is a factor of each term of the right side of the preceding equation, and hence is a factor of $f(x) - f(p)$. Therefore $f(x) - f(p) = (x - p)Q(x)$. The fact that Q is a polynomial should be evident to the reader, and is not proved here.

QED

Corollary 1.19. If f is a polynomial and p is any real number, then $f(x) = (x - p)Q(x) + f(p)$; that is, if $f(x)$ is divided by $x - p$, the remainder is $f(p)$.

Theorem 1.20. If $\langle p_i \rangle$ is any convergent sequence, and f is any polynomial, then the sequence $\langle f(p_i) \rangle$ is bounded.

Proof. Let $f(x) = a_0x^n + a_1x^{n-1} + a_2x^{n-2} + \ldots + a_{n-1}x + a_n$. Then for each $i \in I_o$, we have

$$f(p_i) = a_0(p_i)^n + a_1(p_i)^{n-1} + a_2(p_i)^{n-2} + \ldots + a_{n-1}(p_i) + a_n$$

Since $\langle p_i \rangle$ is a convergent sequence, it is bounded; thus there exists a positive number K such that $|p_i| < K$ for every $i \in I_o$. Using properties of absolute value, we thus have

$$\begin{aligned} |f(p_i)| &= |a_0(p_i)^n + a_1(p_i)^{n-1} + a_2(p_i)^{n-2} + \ldots + a_{n-1}(p_i) + a_n| \\ &\leq |a_0(p_i)^n| + |a_1(p_i)^{n-1}| + |a_2(p_i)^{n-2}| + \ldots + |a_{n-1}(p_i)| + |a_n| \\ &= |a_0| \cdot |p_i|^n + |a_1| \cdot |p_i|^{n-1} + |a_2| \cdot |p_i|^{n-2} + \ldots \\ &\qquad + |a_{n-1}| \cdot |p_i| + |a_n| \\ &< |a_0| \cdot K^n + |a_1| \cdot K^{n-1} + |a_2| \cdot K^{n-2} + \ldots + |a_{n-1}| \cdot K + |a_n| \end{aligned}$$

The last expression is a finite sum of products of real numbers, and hence is a real number M. Then we have $|f(p_i)| < M$ for every $i \in I_o$. But this means that the sequence $\langle f(p_i) \rangle$ is bounded. QED

Alternate proof of Theorem 1.16. Let f be any polynomial, let p be any real number, and let $\langle p_n \rangle$ be any sequence of points in R_1 such that $\lim p_n = p$. We want to show that $\lim f(p_n) = f(p)$. Let $\epsilon > 0$ be given. By Theorem 1.18, for each $n \in I_o$, $f(p) - f(p_n) = (p - p_n) \cdot Q(p_n)$, where Q is a polynomial. By Theorem 1.20, $\langle Q(p_n) \rangle$ is bounded; that is, there exists a real number $K > 0$ such that $|Q(p_n)| < K$ for every $n \in I_o$. Since $\lim p_n = p$, there exists a positive integer N such that for all $n > N$, $|p - p_n| < \epsilon/K$. Thus, for all $n > N$, we have

$$\begin{aligned} |f(p) - f(p_n)| &= |(p - p_n)Q(p_n)| \\ &= |p - p_n| \cdot |Q(p_n)| < \left(\frac{\epsilon}{K}\right) \cdot K = \epsilon \end{aligned}$$

But this says that $\lim f(p_n) = f(p)$. Therefore f is continuous at p. QED

It might be interesting to see how continuity of a particular polynomial can be shown. In the following example, we consider a given polynomial of degree two, and prove continuity in three different ways. The first method relies on the (ϵ, N) definition of a convergent sequence, and shows how, given $\epsilon > 0$, we may find the corresponding value N by means of a "double squeeze." The second method employs the various properties of limits. The third method is a variation of the first, and uses boundedness of a convergent sequence.

Example 1.21. Let $f(x) = x^2 - 3x + 1$, let p be any real number, and let $\langle p_n \rangle$ be any sequence of points in R_1 such that $\lim p_n = p$. We want to show that $\lim f(p_n) = f(p)$. By definition of f, we have $f(p) = p^2 - 3p + 1$, and for each $n \in I_o$, $f(p_n) = p_n{}^2 - 3p_n + 1$. Then

$$\begin{aligned} f(p_n) - f(p) &= (p_n{}^2 - 3p_n + 1) - (p^2 - 3p + 1) \\ &= (p_n{}^2 - p^2) - 3p_n + 3p \\ &= [(p_n - p)^2 + 2pp_n - 2p^2] - 3p_n + 3p \\ &= (p_n - p)^2 + 2p(p_n - p) - 3(p_n - p) \\ &= (p_n - p)^2 + (2p - 3)(p_n - p) \qquad \text{for each } n \in I_o \end{aligned}$$

Since equality holds at each step in the preceding chain, we have

$$(1) \quad f(p_n) - f(p) = (p_n - p)^2 + (2p - 3)(p_n - p) \qquad \text{for each } n \in I_o$$

We are now ready to prove continuity of f at the point p.

First Method. Let $\epsilon > 0$ be given. Then our problem is to find a positive integer N such that $|f(p_n) - f(p)| < \epsilon$ for all $n > N$. Using (1) and the properties of absolute value, we have for each $n \in I_o$,

$$\begin{aligned} |f(p_n) - f(p)| &= |(p_n - p)^2 + (2p - 3)(p_n - p)| \\ &\leq |(p_n - p)^2| + |(2p - 3)(p_n - p)| \\ &= |p_n - p|^2 + |2p - 3| \cdot |p_n - p| \\ &= |p_n - p|(|p_n - p| + |2p - 3|) \end{aligned}$$

We now take advantage of the fact that $\lim p_n = p$ to "put the squeeze on" $|p_n - p|$. We do this in two stages, beginning with the expression in parentheses. Choose the positive integer N_1 such that for all $n > N_1$, $|p_n - p| < 1$. Then, for all $n > N_1$, we have $|p_n - p| + |2p - 3| < 1 + |2p - 3|$. Since p is a constant, it follows that $1 + |2p - 3|$ is a positive constant. We now squeeze $|p_n - p|$ again. Choose the positive integer N_2 such that for all $n > N_2$, $|p_n - p| < \epsilon/(1 + |2p - 3|)$. Now define $N = N_1 + N_2$. Finally, for all $n > N$,

$$\begin{aligned} |f(p_n) - f(p)| &\leq |p_n - p|(|p_n - p| + |2p - 3|) \\ &< |p_n - p|(1 + |2p - 3|) \qquad \text{since } N > N_1 \\ &< \frac{\epsilon}{1 + |2p - 3|}(1 + |2p - 3|) \qquad \text{since } N > N_2 \\ &= \epsilon \end{aligned}$$

Therefore, $\lim f(p_n) = f(p)$.

Second Method. Transposing $f(p)$ to the right side of (1), we obtain $f(p_n) = f(p) + (p_n - p)[(p_n - p) + (2p - 3)]$ for each

$n \in I_o$. We can see (using various properties of limits) that the limit of the right side of the preceding equality exists, and hence $\lim f(p_n)$ exists. We may use here the fact that if $\lim p_n = p$, then $\lim (p_n - p) = 0$. Formalizing this procedure, and using the properties of limits, we have

$$\begin{aligned}\lim f(p_n) &= \lim \{f(p) + (p_n - p)[(p_n - p) + (2p - 3)]\} \\ &= f(p) + \lim (p_n - p)\cdot\lim [(p_n - p) + (2p - 3)] \\ &= f(p) + 0\cdot(0 + 2p - 3) \\ &= f(p)\end{aligned}$$

Therefore, $\lim f(p_n) = f(p)$.

Third Method. Let $\epsilon > 0$ be given. We want to find a positive integer N such that $|f(p_n) - f(p)| < \epsilon$ for all $n > N$. Since the sequence $\langle p_n \rangle$ is convergent, it is bounded; hence, there exists a real number $K_1 > 0$ such that $|p_n| < K_1$ for every $n \in I_o$. Now choose $K > \sup \{K_1, |p|, 3\}$. Then, $K > |p|$, $K > 3$, and $K > |p_n|$ for every $n \in I_o$. Using (1), the properties of absolute value, and the bound K, we have, for every $n \in I_o$

$$\begin{aligned}|f(p_n) - f(p)| &= |(p_n - p)^2 + (2p - 3)(p_n - p)| \\ &= |(p_n - p)(p_n - p + 2p - 3)| \\ &= |p_n - p|\cdot|p_n + p - 3| \\ &\le |p_n - p|\cdot(|p_n| + |p| + 3) \\ &< |p_n - p|\cdot(K + K + K) \\ &= |p_n - p|\cdot 3K\end{aligned}$$

We now use the fact that $\lim p_n = p$. Since $\frac{\epsilon}{3K} > 0$, there exists a positive integer N such that $|p_n - p| < \frac{\epsilon}{3K}$ for all $n > N$. Thus, for all $n > N$, we have

$$|f(p_n) - f(p)| < |p_n - p|\cdot 3K < \frac{\epsilon}{3K}\cdot(3K) = \epsilon$$

This concludes Example 1.21.

Since all polynomials are continuous, it follows that sums, products, and differences of polynomials are continuous; in fact, they are polynomials. It is worth noting that the role of polynomials among all functions is similar to that of the integers among all real numbers. Note, for example, that sums, products, and differences of integers are integers. The analogy may be carried even further. Just as a rational number is defined as the quotient of two integers (with nonzero denominator), we define a *rational function* as the quotient of two polynomials. Of course, a rational function fails to be defined for any value that makes the denominator zero. It follows, however, from Theorem 1.13 that a rational function is continuous wherever it is defined.

In our study of calculus, we considered many functions other than polynomials and rational functions. These were classed generally as *elementary functions*, and included (in addition to polynomials and rational functions) trigonometric, inverse trigonometric, logarithmic, exponential, and various root functions. These elementary functions generally have the property of being continuous wherever they are defined, although the proof of continuity is by no means trivial in most cases. Fortunately, we have been able to establish the continuity of these functions in calculus as a dividend of differentiability (recall that if a function is differentiable at a point p, it is continuous at p). Since we do not plan a discussion of derivatives or integrals in this course, however, we will not pursue the problem of proving continuity of all the elementary functions; rather we accept the fact of their continuity as proved in calculus.

Note the interesting pattern of growth for the collection of continuous functions. As soon as we determine a continuous function, we can combine it with other continuous functions (using the sum, product, difference, and quotient theorems) to construct additional continuous functions. We may also consider the composite function of two continuous functions, and wonder whether or not it is continuous. This question is answered affirmatively in the following theorem.

Theorem 1.22. Let A, B, and C be sets in R_1, let f and g be functions such that $f: A \to B$ and $g: B \to C$, and let h be the composite function gf; so that $h: A \to C$. If f and g are continuous, then h is continuous.

Proof. Let p be any point in A, and let $\langle p_n \rangle$ be a sequence of points in A such that $\lim p_n = p$. We want to show that $\lim h(p_n) = h(p)$. Since f is continuous, we know that $\lim f(p_n) = f(p)$; that is, $\langle f(p_n) \rangle$ is a sequence of points in B which converges to the point $f(p)$ in B. Then, since g is continuous, we have $\lim g[f(p_n)] = g[f(p)]$. Thus, by definition of a composite function, $\lim gf(p_n) = gf(p)$, or $\lim h(p_n) = h(p)$. QED

Now that we have determined a huge collection of continuous functions, we concern ourselves with some of the important properties of continuity. The serious student of calculus should already be aware of a few such properties, although he may not immediately recognize them here in our more general approach. Certainly, even the definition of a continuous function (as one that preserves convergent sequences) should emphasize that we are dealing with a very strong property.

Our basic approach is to determine those properties of sets which are preserved by continuous functions. After the buildup in the preceding paragraph, one might vainly hope that *all* properties of sets are preserved. Unfortunately, since this is not true, we may as well absorb the disappointments first.

Bounded sets are not necessarily preserved by continuous functions, as shown by the following example.

Example 1.23. Let $A = (0, 1)$, and let f be defined by $f(x) = 1/x$ for each $x \in A$. Then it is evident that A is bounded. Also, f is continuous, since it is the quotient of two continuous functions where the denominator function is nonzero (since $0 \notin A$). But the range of f is not bounded.

Open sets are not necessarily preserved by continuous functions, as shown by the following example.

Example 1.24. Let $A = (0, 1)$, and let f be defined by $f(x) = 2$ for every $x \in A$. Then, A is open, and f is continuous, but the range of f is the set $\{2\}$, which is not open.

Closed sets are not necessarily preserved by continuous functions, as shown by the following example.

Example 1.25. Let A be the closed ray $\{x : x \geq 1\}$, and let f be defined by $f(x) = 1/x$ for each $x \in A$. Then A is closed, and f is continuous. The range of f is the set $(0, 1]$, which is not closed. Note that zero is a cluster point of the sequence $\langle 1/n \rangle$, but zero is not in the range of f. Thus, f is a continuous function whose domain is closed, but whose range is not closed.

We are finally ready to obtain some positive results. The following theorem is one of the most important in analysis.

Theorem 1.26. Each of the following properties of a subset H of R_1 is preserved by all continuous mappings.

(*a*) H is compact.

(*b*) H is sequentially compact.

(*c*) H is both closed and bounded.

Proof. By virtue of Theorem 4.23 and Corollary 4.24 of Chapter IV, we see that the three properties just listed are equivalent. Thus our proof will be complete if we show that any one of these properties is preserved under all continuous mappings. We shall prove that the property of sequential compactness is thus preserved.

Let f be a continuous mapping whose domain is H and whose range is K. We want to prove that if H is sequentially compact, K is sequentially compact, and will do so by the contrapositive technique. Suppose K is not sequentially compact. Then, by Problem 57 of Chapter IV, there exists a sequence $\langle q_n \rangle$ of distinct points of K that has no cluster point in K. In particular, no subsequence of $\langle q_n \rangle$ converges to a point of K. Since f is a mapping of H onto K, every point of K is the image under f of at least one point of H. Thus for each point q_n in K, we can choose a point p_n in H such that $f(p_n) = q_n$. Since $\langle q_n \rangle$ is a sequence of distinct points of K, $\langle p_n \rangle$ must be a sequence of distinct points of H. Now let us assume that the sequence $\langle p_n \rangle$ contains a subsequence $\langle p_{n_i} \rangle$ which converges to a point p in H. Then, by the continuity of f, the sequence $\langle f(p_{n_i}) \rangle$ converges to the point $f(p)$ in K. However, for each n_i, $f(p_{n_i}) = q_{n_i}$. Hence $\langle q_{n_i} \rangle$ is a subsequence of $\langle q_n \rangle$ which converges to the point $f(p)$ of K. This is impossible, since we have already shown that no subsequence of $\langle q_n \rangle$ converges to a point of K. We thus conclude that our previous assumption is incorrect; consequently the sequence $\langle p_n \rangle$ contains no subsequence converging to a point of H. Thus $\langle p_n \rangle$ is a sequence of distinct points of H that has no cluster point in H. Therefore, by Problem 57 of Chapter IV, H is not sequentially compact. QED

We thus see that although continuous functions do not necessarily preserve sets that are bounded or sets that are closed, they do preserve sets that are both bounded and closed. It is interesting to glance back at two of our previous examples in the light of this information. Note that the set defined in Example 1.23 is bounded but not closed, and the set defined in Example 1.25 is closed but not bounded.

We now find it convenient to introduce some special terminology for sets which are quite closely related to the range and domain of a mapping.

Definition 1.27. (*a*) If f is a mapping whose domain is the set A, we denote the range of f by $f(A)$; that is,

$$f(A) = \{f(x) : x \in A\}$$

(*b*) Let $f: A \to R_1$ be a mapping. For any subset B of R_1, we define

$$f^{-1}(B) = \{x \in A : f(x) \in B\}$$

The subset $f^{-1}(B)$ of A is called the *source* (or *inverse*) of the set B under the mapping f. We note in particular that if H is any subset of R_1 such that $H \cap f(A) = \phi$, then $f^{-1}(H) = \phi$.

Example 1.28. Let $f: R_1 \to R_1$ be the function defined by $f(x) = x^2 + 1$ for all $x \in R_1$. The range of f is the closed ray $\{x : x \geq 1\}$. Let $B_1 = [1, 5]$ be a subset of R_1, and note that B_1 is a subset of the range $f(R_1)$ of f. We want to determine the set $f^{-1}(B_1)$. Now, $f(x) \in B_1$ iff $1 \leq f(x) \leq 5$, or iff $1 \leq x^2 + 1 \leq 5$, or iff $0 \leq x^2 \leq 4$, or iff $|x| \leq 2$. Thus, we see that $f^{-1}(B_1) = [-2, 2]$. We next consider the subset B_2 of R_1 defined by $B_2 = [-10, 5]$, and note that B_2 is not a subset of the range $f(R_1)$ of f. Observe that for every $x \in R_1$, $f(x) \in \{x : x \geq 1\}$. Making use of this fact, we see that a point x is in $f^{-1}(B_2)$ iff $f(x)$ is in both the closed ray $\{x : x \geq 1\}$ and the closed interval $[-10, 5]$, that is, iff

$$f(x) \in \{x : x \geq 1\} \cap [-10, 5] = B_1$$

Therefore $f^{-1}(B_2) = f^{-1}(B_1) = [-2, 2]$.

The discussion in Example 1.28 leads at once to the following theorem, the proof of which is left as an exercise.

Theorem 1.29. Let $f: A \to R_1$ be a mapping, and B any subset of R_1. Then

$$f^{-1}(B) = f^{-1}[B \cap f(A)]$$

Theorem 1.30. Let A be any closed and bounded set in R_1, and let f be any continuous function from A into R_1. Then there exist points x_1 and x_2 in A such that

$$f(x_1) \leq f(x) \leq f(x_2)$$

for every x in A. In other words, f is bounded on A and takes on both its supremum and its infimum on A.

Proof. Since A is closed and bounded, it follows from Theorem 1.26 that $f(A)$ is closed and bounded. Hence by Theorem 4.10 of Chapter IV, $f(A)$ contains both its infimum and its supremum. Let us denote these by m and M, respectively, so that for every x in A, we have $m \leq f(x) \leq M$. But f maps A onto $f(A)$, so there exist points x_1 and x_2 in A such that $f(x_1) = m$ and $f(x_2) = M$.

QED

Since a closed interval is both closed and bounded, we have the following corollary.

Corollary 1.31. Let f be any continuous function defined on the closed interval $A = [a, b]$. Then $f(A)$ contains its least upper bound M and its greatest lower bound m; that is, there exist points x_1 and x_2 in $[a, b]$ such that

$$f(x_1) = m, f(x_2) = M, \qquad \text{and} \qquad f(x_1) \leq f(x) \leq f(x_2)$$

for every x in $[a, b]$.

The statement of this corollary should be familiar to us from our work in calculus, although most elementary texts omit its proof. This property of continuity is often stated in the form: "A continuous function on a closed interval attains a maximum value and a minimum value on the interval," from which it follows as a corollary that "A continuous function on a closed interval is bounded."

One important class of sets remains, namely, connected sets, whose behavior under continuous mappings should be investigated. We shall presently show that connected sets are preserved by continuous functions. To do this, we make use of two lemmas to obtain a characterization of connected subsets of R_1.

Lemma 1.32. Let H be a set in R_1, and suppose that a and b are distinct points of R_1 and that $f: H \xrightarrow{\text{onto}} \{a\} \cup \{b\}$ is continuous. Let $A = f^{-1}(\{a\})$ and $B = f^{-1}(\{b\})$. Then

$$\overline{A} \cap B = \phi \qquad \text{and} \qquad A \cap \overline{B} = \phi$$

Proof. It is evident from the definitions of A and B that $H = A \cup B$ and $A \cap B = \phi$. Thus if $p \in \overline{A} \cap B$, p is a cluster point of A. Hence there exists a sequence $\langle p_n \rangle$ of points of A such that $\lim p_n = p$. Since f is continuous, it follows that $\lim f(p_n) = f(p)$. But this is impossible, since $f(p_n) = a$ for every n (and hence $\lim f(p_n) = a$), whereas $f(p) = b$. Therefore $\overline{A} \cap B = \phi$. A similar proof shows that $A \cap \overline{B} = \phi$.

Lemma 1.33. Under the hypotheses of Lemma 1.32, the set H is not connected.

Proof. We see that $H = A \cup B$, $A \neq \phi$, $B \neq \phi$. Let $p \in A$ and $q \in B$. Since $A \cap B = \phi$, it follows that $p \neq q$, and we may choose our notation so that $p < q$. Assume that H is connected. Then $[p, q] \subset H$. Define the set L as follows: $L = [p, q] \cap B$. The set L is not empty, since $q \in L$; furthermore, the point p is a lower bound for L. Thus L has an infimum; so let us define $h = \inf L$. If $h \in L$, then $h \in \overline{L}$ by (3) of Theorem 4.17 of Chapter IV. If $h \notin L$, then by Problem 59 of Chapter IV, h is a cluster point of L, so $h \in \overline{L}$. Thus, in either case, we have $h \in \overline{L}$. Using various properties of intersection and closure, we see that

$$\begin{aligned} \overline{L} &= \overline{[p, q] \cap B} \\ &\subset \overline{[p, q]} \cap \overline{B} \qquad \text{by (6), Theorem 4.17 of Chapter IV} \\ &\subset \overline{B} \end{aligned}$$

Therefore $h \in \overline{B}$; hence $h \notin A$, since $A \cap \overline{B} = \phi$ by Lemma 1.32. But $H = A \cup B$, and since $h \in H$, we must have $h \in B$. Thus, $p < h$. Since $[p, q] \subset A \cup B$, where $A \cup B$ is connected, and since no point between p and h can belong to B, we see that the open interval $(p, h) \subset A$, so that $h \in \overline{A}$. Thus, $h \in \overline{A} \cap B$. But this contradicts Lemma 1.32. Therefore H is not connected.

QED

Theorem 1.34. The set H in R_1 is not connected iff there exist two distinct points a and b of R_1 and a continuous function $f: H \xrightarrow{\text{onto}} \{a\} \cup \{b\}$.

Proof. We must prove the following two statements:

(i) If there exist two distinct points a and b of R_1 and a continuous function $f: H \xrightarrow{\text{onto}} \{a\} \cup \{b\}$, H is not connected.

(ii) If H is not connected, there exist two distinct points a and b of R_1 and a continuous function $f: H \xrightarrow{\text{onto}} \{a\} \cup \{b\}$.

We have already proved (i) in Lemma 1.33. To prove (ii) suppose that H is not connected. Then, there exist points $p, q \in H$ and $r \notin H$ such that r is between p and q. We may choose our notation so that $p < q$. Define $A = \{x \in H: x < r\}$ and $B = \{x \in H: x > r\}$. Note that neither A nor B is empty, since $p \in A$ and $q \in B$. Now let a and b be any two distinct points of R_1 and define the function $f: H \xrightarrow{\text{onto}} \{a\} \cup \{b\}$ as follows:

$$f(x) = \begin{cases} a & \text{if } x \in A \\ b & \text{if } x \in B \end{cases}$$

We leave as an exercise the proof that f is continuous. QED

Theorem 1.35. Connected sets in R_1 are preserved by continuous functions.

Proof. Let H be a connected set in R_1 and $f: H \to R_1$ a continuous function. We want to show that $f(H)$ is connected. Let $K = f(H)$. We prove the contrapositive of the theorem by showing that if K is not connected, H is not connected. Thus, suppose K is not connected. By Theorem 1.34, there exist two distinct points a and b of R_1 and a continuous function $g: K \xrightarrow{\text{onto}} \{a\} \cup \{b\}$. Define the function $h: H \to R_1$ as follows: For every $x \in H$, $h(x) = g[f(x)]$. It is easy to see that $h = gf: H \xrightarrow{\text{onto}} \{a\} \cup \{b\}$. However, h is the composite function of two continuous functions, and hence is continuous. Therefore H is not connected, by Theorem 1.34. QED

One of the most frequently used theorems of analysis is a corollary to Theorem 1.35.

Corollary 1.36 (The Intermediate Value Theorem). Let f be defined and continuous on a connected set H of R_1. Let x_1 and x_2, $x_1 < x_2$ be two points of H such that $f(x_1) \neq f(x_2)$ and let c be any real number between $f(x_1)$ and $f(x_2)$. Then there exists a point x_3 of H such that $x_1 < x_3 < x_2$ and $f(x_3) = c$.

This theorem is often stated in calculus in the following general form: "A function continuous on an interval assumes as a value (at least once) every number between any two of its values." The student first employed the intermediate value theorem in college algebra when he made use of the fact that if f is a polynomial such that $f(a)$ and $f(b)$ have opposite signs, then the equation $f(x) = 0$ has a real root between a and b.

PROBLEMS

1. Prove Theorem 1.8.
2. Prove Theorem 1.11.
3. Prove Theorem 1.13.
4. Give examples to show that the sum, product, difference, and quotient of two functions may be continuous, even though the two given functions are not continuous.
5. Prove Lemma 1.17. [*Hint:* Use mathematical induction, and the identity $x^{n+1} - p^{n+1} = x(x^n - p^n) + p^n(x - p)$.]
6. Let f be a polynomial, and p a real number. Prove that $(x - p)$ is a factor of $f(x)$ iff $f(p) = 0$.
7. Give an (ϵ, N) proof of each of the following, where $\lim a_n = A$.
   (*a*) $\lim f(a_n) = 2A^2 - 3A + 1$, where $f(x) = 2x^2 - 3x + 1$.
   (*b*) $\lim f(a_n) = 5A^2 + 4A - 17$, where $f(x) = 5x^2 + 4x - 17$.
   (*c*) $\lim f(a_n) = A^3 - 4A^2 + 3A - 1$, where $f(x) = x^3 - 4x^2 + 3x - 1$.
8. Determine the range of the function f defined in Example 1.23, and show that this range is unbounded.
9. Prove Theorem 1.29.
10. In the proof of Theorem 1.26, show that the sequence $\langle p_n \rangle$ must be a sequence of distinct points of H. Show how this sequence may be found using the Axiom of Choice for Sequences.
11. In Lemma 1.32, prove that $A \cap \overline{B} = \phi$.
12. Complete the proof of Theorem 1.34 by showing that the function f is continuous.
13. Show that the equation $x^6 + 6x - 5 = 0$ has a real root.
14. Give an example of a function which is bounded on the closed interval $[0, 1]$, but which attains neither a maximum value nor a minimum value on the interval.
15. Give an example of a function which is defined on the closed interval $[0, 1]$, but which is not bounded.
16. Give an example of a one-to-one continuous mapping of the open interval $(0, 1)$ onto R_1.
17. Let A and B be subsets of R_1 and f a function such that $f: A \to B$. If q is a real number not in B, show that $f^{-1}(\{q\}) = \phi$.
18. Let W and B be subsets of R_1, and define $K = \overline{W} \cap B$. Prove that $K = \overline{K} \cap B$.
19. Let $f: A \to R_1$ be a mapping, and let B and C be subsets of A. Prove
   (*a*) $f(B \cup C) = f(B) \cup f(C)$.

(*b*) $f(B \cap C) \subset f(B) \cap f(C)$.

(*c*) Give an example in which $f(B \cap C) \neq f(B) \cap f(C)$.

20. Let $f: A \to R_1$ be a mapping, and define $B = f(A)$.

Suppose that C and D are subsets of R_1. Prove that

(*a*) $f^{-1}(C \cup D) = f^{-1}(C) \cup f^{-1}(D)$.

(*b*) $f^{-1}(C \cap D) = f^{-1}(C) \cap f^{-1}(D)$.

(*c*) $f[f^{-1}(C)] = C \cap B$.

(*d*) For any subset E of A, $f^{-1}[f(E)] \supset E$.

(*e*) Give an example of a mapping $f: A \to R_1$ and a subset E of A such that $f^{-1}[f(E)] \neq E$.

21. Let $f: A \to R_1$ be a mapping. State and prove an "iff" condition on f which will insure that $f^{-1}[f(E)] = E$ for every subset E of A.

22. Let $f: A \to B$ be a mapping, where A and B are subsets of R_1, and let K be any subset of B. Prove that $f^{-1}[C(K)] = C[f^{-1}(K)]$.

23. Let $J = [0, 1]$, and let $f: J \to J$ be any continuous mapping. Prove that there exists a point $x_0 \in J$ such that $f(x_0) = x_0$.

(*Hint:* Consider the mapping $g: R_1 \to R_1$ defined by $g(x) = f(x) - x$ for $x \in R_1$).

2. *Other criteria for continuity*

We have defined continuous functions as those functions which preserve convergent sequences in R_1. Sometimes, however, it is neither practicable nor desirable to work with sequences when dealing with continuity of a function. We thus seek additional criteria for continuity—criteria which do not depend on the notion of convergent sequences. In this section, we derive some equivalent formulations for continuity, and do so through the use of the concept of a suburb of a point.

Definition 2.1. A set K in R_1 is a *suburb* of a point p in R_1 iff $C(K)$ is a neighborhood of p in R_1.

The following theorem is an immediate consequence of this definition, and its proof is left as an exercise.

Theorem 2.2. A set K in R_1 is a suburb of a point p in R_1 iff K is a closed set in R_1 which does not contain p; that is, iff $K = \bar{K}$ and $p \notin \bar{K}$.

We may now express convergence of a sequence in terms of suburbs.

Theorem 2.3. The sequence $\langle p_n \rangle$ of points in R_1 converges to the point p in R_1 iff no suburb of p contains a subsequence of $\langle p_n \rangle$.

Proof. We must prove the following two statements:

(1) If $\lim p_n = p$, then no suburb of p contains a subsequence of $\langle p_n \rangle$.

(2) If no suburb of p contains a subsequence of $\langle p_n \rangle$, then $\lim p_n = p$.

To prove (1), suppose that $\lim p_n = p$, and let K be any suburb of p. Then $C(K)$ is a neighborhood of p, and hence $C(K)$ contains all but at most a finite number of terms of $\langle p_n \rangle$; in particular, K does not contain a subsequence of $\langle p_n \rangle$. This proves (1). To prove (2), suppose that no suburb of p contains a subsequence of $\langle p_n \rangle$, and let G be any neighborhood of p. Then $C(G)$ is a suburb of p, and hence $C(G)$ contains no subsequence of $\langle p_n \rangle$. Thus $C(G)$ contains at most a finite number of points of $\langle p_n \rangle$, and consequently G contains all but at most a finite number of points of $\langle p_n \rangle$. Therefore $\lim p_n = p$. QED

Since the contrapositive form of Theorem 2.3 will be useful to us, we now state it as a corollary.

Corollary 2.4. The sequence $\langle p_n \rangle$ of points in R_1 does not converge to the point p in R_1 iff there exists a suburb of p which contains a subsequence of $\langle p_n \rangle$.

Lemma 2.5. Let A and B be sets in R_1 and $f: A \to B$. If f is continuous at the point p of A, and if K is any suburb of $f(p)$, then $f^{-1}(K)$ is a suburb of p.

Proof. Let $H = f^{-1}(K)$, so that $f(H) \subset K$. Suppose $\overline{H}$ is not a suburb of p. Then $p \in \overline{H}$, and hence there exists a sequence $\langle p_n \rangle$ of points of H converging to the point p. Since f is continuous at p by hypothesis, we see that $\langle f(p_n) \rangle$ converges to the point $f(p)$ in B. But $f(H) \subset K$, and hence the sequence $\langle f(p_n) \rangle$ is contained in K. This is impossible, by Theorem 2.3, since K is a suburb of $f(p)$. QED

Lemma 2.6. Let A and B be sets in R_1 and $f: A \to B$. Let p be any point of A. If for every suburb K of $f(p)$, $\overline{f^{-1}(K)}$ is a suburb of p, then f is continuous at p.

Proof. Let $\langle p_n \rangle$ be any sequence of points of A such that $\lim p_n = p$. We want to show that $\lim f(p_n) = f(p)$. Suppose that

$\langle f(p_n)\rangle$ does not converge to $f(p)$. Then by Corollary 2.4, there exists a suburb K of $f(p)$ which contains a subsequence $\langle f(p_{n_i})\rangle$ of $\langle f(p_n)\rangle$. Therefore $\overline{f^{-1}(K)}$ contains the subsequence $\langle p_{n_i}\rangle$ of $\langle p_n\rangle$, and hence $\overline{f^{-1}(K)}$ is not a suburb of p, by Theorem 2.3. QED

Combining the results of Lemmas 2.5 and 2.6, we have the following.

Theorem 2.7. Let A and B be sets in R_1 and $f: A \to B$. Let p be any point in A. Then f is continuous at p iff for every suburb K of $f(p)$, $\overline{f^{-1}(K)}$ is a suburb of p.

Before continuing our discussion of continuity, we introduce some additional notation and terminology.

Definition 2.8. Let A and B be sets in a universe U. The notation $A - B$ is used to designate the set of all points of U which belong to A but do not belong to B; that is, $A - B = A \cap C(B)$. We sometimes refer to the set $A - B$ as the *complement of the set B with respect to the set A.*

Note that if, in Definition 2.8, the set A is taken to be the universe U, we then have $U - B = U \cap C(B) = C(B)$, so that $U - B$ is an alternate form for expressing the complement of the set B with respect to the universe U.

Definition 2.9. Let A and K be subsets of R_1. Then
(*a*) K is said to be *closed in A* iff $K = \overline{K} \cap A$.
(*b*) K is said to be *open in A* iff $A - K$ is closed in A.

Our next theorem is the most important characterization of continuous mappings in analysis. Its importance lies in the fact that it ties together so beautifully the concepts of "open sets," "closed sets," and "continuous mappings."

Theorem 2.10. Let A and B be subsets of R_1, and $f: A \to B$ a mapping. Then f is continuous iff either of the following conditions holds:

(1) If K is any subset of B which is closed in B, then $f^{-1}(K)$ is closed in A.

(2) If K is any subset of B which is open in B, then $f^{-1}(K)$ is open in A.

The statement of this theorem means that if f is continuous, then both (1) and (2) are true; and if either (1) or (2) is true, then

f is continuous. Thus the proof of Theorem 2.10 will consist of a number of lemmas.

Lemma 2.11. If f is continuous, then (1) is true.

Proof. We shall use the contrapositive technique. Suppose that K is a subset of B which is closed in B, but that the set $H = f^{-1}(K)$ is not closed in A. We know that $H \subset \overline{H}$ and $H \subset A$. Thus $H \subset \overline{H} \cap A$. However, since H is not closed in A, we know that $H \neq \overline{H} \cap A$. This means that there exists a point p such that $p \in \overline{H} \cap A$, but $p \notin H$. In particular, $p \in \overline{H}$.

We note that $f(H) = K \cap f(A)$. Since $p \notin H$, and since $H = f^{-1}(K)$, $f(p) \notin K = \overline{K} \cap B$. However, $f(p) \in B$, and hence $f(p) \notin \overline{K}$. Therefore $\overline{K}$ is a suburb of $f(p)$. Consequently, as a result of the continuity of f and Theorem 2.7, it follows that $\overline{H}$ is a suburb of p. But this is impossible, since $p \in \overline{H}$. QED

Lemma 2.12. If (1) is true, then f is continuous.

Proof. The statement "f is continuous" means that "given any point p of A, then f is continuous at p." Thus let p be any point of A, and let W be any suburb of $f(p)$. Define $K = \overline{W} \cap B$. We want to show that K is closed in B, so we must prove that $K = \overline{K} \cap B$. This is immediate from Problem 18. Define $H = f^{-1}(K)$. Since (1) is true, $H = \overline{H} \cap A$.

We now want to show that $\overline{H}$ is a suburb of p. To do this, we note that W is a suburb of $f(p)$, so that $f(p) \notin \overline{W}$. It then follows from the definition of K that $f(p) \notin K$. Since $H = f^{-1}(K)$, we thus have $p \notin H$. But $p \in A$ by our original choice of p. Using the fact that $H = \overline{H} \cap A$, we therefore have $p \notin \overline{H}$, so that $\overline{H}$ is a suburb of p. Finally, we must show that $\overline{f^{-1}(W)}$ is a suburb of p. Since W is a suburb of $f(p)$, we know that $W = \overline{W}$, by Theorem 2.2, and hence $K = \overline{W} \cap B = W \cap B$, so that

$$\overline{H} = \overline{f^{-1}(K)} = \overline{f^{-1}(W \cap B)} = \overline{f^{-1}(W)}$$

by Theorem 1.29; so $\overline{f^{-1}(W)}$ is a suburb of p. Therefore f is continuous at p, by Lemma 2.6. QED

We have now reached the point in the proof of Theorem 2.10 where we have shown that f is continuous iff (1) is true. The completion of the proof of the theorem consists of the following two lemmas, the first of which is left as an exercise.

Lemma 2.13. Let A and B be sets in R_1, and $f: A \to B$ a mapping. If K is any subset of B, then $f^{-1}(B - K) = f^{-1}(B) - f^{-1}(K)$.

Lemma 2.14. f is continuous iff (2) holds.

Proof. We must prove the following two statements:

(*a*) If f is continuous, then given any subset K of B which is open in B, the set $f^{-1}(K)$ is open in A.

(*b*) If given any subset K of B which is open in B, the set $f^{-1}(K)$ is open in A, then f is continuous.

To prove (*a*), suppose f is continuous, and let K be any subset of B which is open in B. Then it follows from Definition 2.9 that $B - K$ is closed in B. Since f is continuous, we know that (1) is true, so $f^{-1}(B - K)$ is closed in A. But by Lemma 2.13, $f^{-1}(B - K) = f^{-1}(B) - f^{-1}(K) = A - f^{-1}(K)$, and hence $A - f^{-1}(K)$ is closed in A. Therefore, by Definition 2.9, $f^{-1}(K)$ is open in A. This proves (*a*).

To prove (*b*), suppose that given any subset K of B which is open in B, the set $f^{-1}(K)$ is open in A. We want to prove that f is continuous. Let F be any subset of B which is closed in B. Then by Definition 2.9, $B - F$ is open in B. From the hypothesis of (*b*), we thus have $f^{-1}(B - F)$ is open in A. But by Lemma 2.13, $f^{-1}(B - F) = f^{-1}(B) - f^{-1}(F) = A - f^{-1}(F)$, and hence $A - f^{-1}(F)$ is open in A, which implies that $f^{-1}(F)$ is closed in A. We thus have shown that if F is any subset of B which is closed in B, then $f^{-1}(F)$ is closed in A. Therefore f is continuous by (1).

This completes the proof of Theorem 2.10.

Still another important equivalent formulation of continuity is expressed in the next theorem. First, however, we list a few lemmas which are used in the proof. The proofs of these lemmas are easy, and are left as exercises.

Lemma 2.15. If $f: A \to R_1$, and if E and F are subsets of A such that $E \subset F$, then $f(E) \subset f(F)$.

Lemma 2.16. Let $f: A \to R_1$ be a mapping, let K be a subset of R_1, and define $H = f^{-1}(K)$. Then $f(A - H) = f(A) - f(H)$.

Lemma 2.17. Let E, F, and G be subsets of R_1. Then

$$E - (F \cap G) = (E - F) \cup (E - G)$$

Theorem 2.18. Let A be a set in R_1, p a point of A, and $f: A \to R_1$ a mapping. Then f is continuous at p iff given any open set G in R_1 such that $f(p) \in G$, there exists an open set V in R_1 such that $p \in V$ and $f(V \cap A) \subset G$.

Proof. We must prove the following two statements:

(1) If f is continuous at p, then given any open set G in R_1 such that $f(p) \in G$, there exists an open set V in R_1 such that $p \in V$ and $f(V \cap A) \subset G$.

(2) If, given any open set G of R_1 such that $f(p) \in G$, there exists an open set V in R_1 such that $p \in V$ and $f(V \cap A) \subset G$, then f is continuous at p.

To prove (1), suppose that f is continuous at p, and let G be any open set in R_1 such that $f(p) \in G$. Define $K = R_1 - G$. Then K is a suburb of $f(p)$. Let $H = f^{-1}(K)$. Since f is continuous at p, it follows from Lemma 2.5 that $\overline{H}$ is a suburb of p. Thus there exists an open set V in R_1 such that $p \in V$ and $V \cap H = \phi$. Therefore $(V \cap A) \cap H = \phi$, and hence $V \cap A \subset C(H)$. But we also have $V \cap A \subset A$; so

$$V \cap A \subset A \cap C(H) = A - H$$

We thus see that

$$\begin{aligned}
f(V \cap A) &\subset f(A - H) && \text{by Lemma 2.15} \\
&= f(A) - f(H) && \text{by Lemma 2.16} \\
&= f(A) - [K \cap f(A)] && \text{since } f(H) = K \cap f(A) \\
&= [f(A) - K] \cup [f(A) - f(A)] && \text{by Lemma 2.17} \\
&= f(A) - K \\
&\subset R_1 - K \\
&= G
\end{aligned}$$

This completes the proof of (1).

To prove (2), suppose the condition of (2) holds. We want to show that f is continuous at p. Let K be any suburb of $f(p)$, and define $G = R_1 - K$. Then G is an open set in R_1 and $f(p) \in G$. Thus by the hypothesis of (2), there exists an open set V in R_1 such that $p \in V$ and $f(V \cap A) \subset G$. Since $p \in V$ and $p \in A$, it follows that $p \in V \cap A$. But $V \cap A \subset f^{-1}(G)$, and $f^{-1}(G) = f^{-1}(R_1 - K) = f^{-1}(R_1) - f^{-1}(K) = A - f^{-1}(K)$, by Lemma 2.13. Therefore $p \in A - f^{-1}(K)$. Now, let us assume that $p \in \overline{f^{-1}(K)}$. We know that $p \notin f^{-1}(K)$; hence p is a cluster point of $f^{-1}(K)$. Since V is a neighborhood of p, it follows that there exists a point q in V such that $q \in f^{-1}(K)$. Thus $q \in V \cap f^{-1}(K)$.

Since A is the domain of f, we must have $f^{-1}(K) \subset A$; consequently $q \in V \cap f^{-1}(K) \cap A$. Then $q \in V \cap A$ and $q \in f^{-1}(K)$. Therefore $f(q) \in f(V \cap A)$ and $f(q) \in K$. But this contradicts the fact that $f(V \cap A) \subset G = R_1 - K$. Since our above assumption leads to a contradiction, we may therefore conclude that $p \notin \overline{f^{-1}(K)}$, and hence $\overline{f^{-1}(K)}$ is a suburb of p. Therefore, by Lemma 2.6, f is continuous at p. QED

As a corollary to Theorem 2.18, we have the following important criterion for continuity. This particular formulation is often used as the definition of continuity in calculus texts.

Theorem 2.19 (The "ϵ, δ Criterion for Continuity"). Let A be a set in R_1, p a point of A, and $f: A \to R_1$. Then f is continuous at p iff given any real number $\epsilon > 0$, there exists a real number $\delta > 0$ such that $|f(x) - f(p)| < \epsilon$ for every x in A such that $|x - p| < \delta$.

Proof. Suppose f is continuous at p, and let $\epsilon > 0$ be given. We must prove the existence of a real number $\delta > 0$ such that $|f(x) - f(p)| < \epsilon$ for every x in A such that $|x - p| < \delta$. Define $G = (f(p) - \epsilon, f(p) + \epsilon)$. Then G is an open set in R_1 such that $f(p) \in G$. We are given that f is continuous at p. Thus by Theorem 2.18, there exists an open set V in R_1 such that $p \in V$ and $f(V \cap A) \subset G$. Since V is open, there exists an open interval (a, b) such that $p \in (a, b) \subset V$. Let $\delta = \inf \{p - a, b - p\}$. Then $(p - \delta, p + \delta) \subset V$, and hence $(p - \delta, p + \delta) \cap A \subset V \cap A$. Thus for every x in A such that $x \in (p - \delta, p + \delta)$, it follows that $f(x) \in G$. Now let x be any point in A such that $|x - p| < \delta$. Then it is easy to see that $x \in (p - \delta, p + \delta)$; hence $f(x) \in G = (f(p) - \epsilon, f(p) + \epsilon)$, which says that $|f(x) - f(p)| < \epsilon$. This proves half the theorem.

To prove the other half, suppose that given any $\epsilon > 0$, there exists $\delta > 0$ such that $|f(x) - f(p)| < \epsilon$ for every x in A such that $|x - p| < \delta$. We want to prove that f is continuous at p. Let G be any open set in R_1 such that $f(p) \in G$. Since G is open, there exists an open interval (a, b) such that $f(p) \in (a, b) \subset G$. Let $\epsilon = \inf \{f(p) - a, b - f(p)\}$. Then, for a given point x in A, it follows that $f(x) \in (f(p) - \epsilon, f(p) + \epsilon)$ iff $|f(x) - f(p)| < \epsilon$. By the condition of the theorem, there exists a real number $\delta > 0$ such that $|f(x) - f(p)| < \epsilon$ for every x in A such that $|x - p| < \delta$.

Define $V = (p - \delta, p + \delta)$. Then $p \in V$. Now let x be any point in $V \cap A$. Then $|x - p| < \delta$, so $|f(x) - f(p)| < \epsilon$, and hence $f(x) \in G$. Thus $f(V \cap A) \subset G$. Therefore f is continuous, by Theorem 2.18. QED

PROBLEMS

24. Prove Theorem 2.2.

25. Let p be a point of R_1. Then prove

(*a*) The union of any finite collection of suburbs of p is a suburb of p.

(*b*) The intersection of any collection of suburbs of p is a suburb of p.

26. Give an example in which the empty set is a suburb of a point.

27. In the proof of Lemma 2.5, prove that there exists a sequence $\langle p_n \rangle$ of points of H converging to the point p.

28. Prove that a point p is in $\overline{H}$ iff there is a sequence of points of H converging to p. If $p \in \overline{H} - H$, prove that there exists a sequence of distinct points of H converging to p.

29. Prove that the union of any finite number of sets, each closed in a set A, is closed in A.

30. Prove that the intersection of any collection of sets, each closed in A, is closed in A.

31. State and prove results similar to Problems 29 and 30 for sets open in A.

32. Prove Lemma 2.13.

33. Prove Lemma 2.15.

34. Prove Lemma 2.16.

35. Prove Lemma 2.17.

36. In the proof of Theorem 2.18, show that there exists an open set V in R_1 such that $p \in V$, and $V \cap H = \phi$.

37. Let $f: R_1 \to R_1$ be a continuous function. Given any subset A of R_1, prove that $f(\overline{A}) \subset \overline{f(A)}$.

38. Let $f: R_1 \to R_1$ be any function such that given any subset A of R_1, $f(\overline{A}) \subset \overline{f(A)}$. Prove that f is continuous.

39. Use the results of Problems 37 and 38 to state an "iff" condition for continuity.

40. Let $f: R_1 \to R_1$ be the mapping defined by $f(x) = 3x - 7$ for $x \in R_1$. Prove that f is continuous at the point $x = 5$ by using the ϵ,δ criterion. For a given $\epsilon > 0$, find an explicit expression for δ. Can you generalize this proof to show that f is continuous at any point p? Does the value of δ in this case depend on the value of p?

41. Repeat Problem 40 for the mapping f defined by $f(x) = x^2$ for each $x \in R_1$.

42. Let A and B be subsets of R_1, and $f: A \xrightarrow{\text{onto}} B$ a continuous mapping with the property that for each point q of B, $f^{-1}(\{q\})$ is a single point p. Suppose that A is compact. Define a mapping $g: B \to A$ as follows: For each $q \in B$, $g(q) = f^{-1}(\{q\})$. Prove that g is continuous.
43. Prove (1) of Theorem 2.18 in the following way: Define the set

$$H = \{x: x \in A, f(x) \notin G\}$$

Suppose that for any neighborhood V of p, $f(V \cap A) \cap C(G) \neq \phi$. Under these conditions, show that p is a cluster point of H. Obtain from this fact a contradiction of the hypothesis that f is continuous at p.

3. Uniform continuity and extension theorems

Before introducing the concept of uniform continuity, let us state precisely what is meant by continuity of a mapping f on a set A. To do this, we reexamine the ϵ,δ Criterion, which we restate here for convenience of reference.

The mapping $f: A \to R_1$ is continuous on the set A iff for each p in A and each $\epsilon > 0$, there exists a real number $\delta > 0$ (depending in general on both p and ϵ) such that $|f(x) - f(p)| < \epsilon$ for every x in A such that $|x - p| < \delta$.

Perhaps the best way to illustrate the dependence of δ on both p and ϵ is by an example.

Example 3.1. Let $A = (0, 1)$ and define $f: A \to R_1$ by $f(x) = 1/x$ for each $x \in A$. Let $p \in A$. We shall show that f is continuous at the point p. To do this, let $\epsilon > 0$ be given. We must find a real number $\delta > 0$ such that $|f(x) - f(p)| < \epsilon$ for every x in A such that $|x - p| < \delta$. We note that for any $x \in A$,

$$|f(x) - f(p)| = \left|\frac{1}{x} - \frac{1}{p}\right| = \frac{|p - x|}{|p| \cdot |x|}$$

Our problem is to make this fraction "small" by choosing x "close to p." This means that we must be sure that the numerator of this fraction is not "too large," and that the denominator is not "too small." We first take care of the denominator. Define $\delta_1 = p/2$. Then

$$|x - p| < \delta_1 \quad \text{iff} \quad p - \frac{p}{2} < x < p + \frac{p}{2} \quad \text{iff} \quad \frac{p}{2} < x < \frac{3p}{2}$$

Thus if $|x - p| < \delta_1$, then $|x| > p/2$, and hence

$$|f(x) - f(p)| < \frac{2|x - p|}{p^2}$$

The problem remains of keeping the numerator "small." How "small" must we make $|x - p|$ in order to be sure that $(2|x - p|)/p^2 < \epsilon$? If we set $(2|x - p|)/p^2 = \epsilon$, we observe that $|x - p| = p^2\epsilon/2$. It follows at once that if $|x - p| < \delta_2 = p^2\epsilon/2$, then $(2|x - p|)/p^2 < \epsilon$. Consequently if $|x - p| < \delta = \inf \{\delta_1, \delta_2\}$, then we have

$$|f(x) - f(p)| < \epsilon$$

Note carefully that the definition of δ can be written as follows:

$$\delta = \inf \left\{ \frac{p}{2}, \frac{p^2\epsilon}{2} \right\}$$

which certainly displays the dependence of δ on both p and ϵ.

Generally, in applying the ϵ,δ Criterion in proofs of continuity, we shall find that the δ which must be chosen will depend both on the preassigned value of ϵ and on the point p at which continuity is to be established. This was indeed true in Example 3.1.

We are extremely fortunate when we are able to find a value of δ which will depend only on the preassigned value of ϵ, and which will be the same for every point p in the domain A of the mapping. It is of the greatest importance that this may always be done when the domain A of the mapping is compact. Before proving this theorem, we introduce the concept of uniform continuity.

Definition 3.2. The mapping $f: A \to R_1$ is said to be *uniformly continuous* iff given any $\epsilon > 0$, there exists a $\delta > 0$ such that if p_1 and p_2 are any two points of A with $|p_1 - p_2| < \delta$, then $|f(p_1) - f(p_2)| < \epsilon$.

Theorem 3.3. Every uniformly continuous function $f: A \to R_1$ is continuous on A.

Proof. Left as an exercise.

We note by Problem 44(*d*) that the mapping f defined in Example 3.1 is not uniformly continuous. On the other hand, the mapping f defined in Problem 40 is uniformly continuous. This is a special case of the following more general theorem.

Theorem 3.4. If $f: R_1 \to R_1$ is defined by $f(x) = ax + b$ for every $x \in R_1$, where $a, b \in R_1$, then f is uniformly continuous.

Proof. Pick any $\epsilon > 0$. If $a \neq 0$, define $\delta = \epsilon/|a|$; if $a = 0$, define $\delta = 1$. Let p_1 and p_2 be any two points of A such that $|p_1 - p_2| < \delta$. We have at once

$$|f(p_1) - f(p_2)| = |a| \cdot |p_1 - p_2| < \epsilon$$

Therefore f is uniformly continuous. QED

Corollary 3.5. The constant function $f: R_1 \to R_1$ defined by $f(x) = b$ for every $x \in R_1$ is uniformly continuous.

Theorem 3.6. Let A be any compact subset of R_1, and $f: A \to R_1$ any continuous mapping. Then f is uniformly continuous.

Proof. Pick any $\epsilon > 0$. For any point $p \in A$, we know that f is continuous at p. By the ϵ, δ Criterion, there must therefore exist a $\delta > 0$, which we denote by δ_p, such that if $|x - p| < 2\delta_p$, then $|f(x) - f(p)| < \epsilon/2$. Define the open set $G_p = (p - \delta_p, p + \delta_p)$. Repeat this for every point $p \in A$. The collection of sets $\{G_p\}$ is an open covering of A. Since A is compact, there exists a finite subcollection

$$\{G_{p_1}, G_{p_2}, G_{p_3}, \ldots, G_{p_n}\}$$

of the collection $\{G_p\}$ such that

$$A \subset G_{p_1} \cup G_{p_2} \cup G_{p_3} \cup \ldots \cup G_{p_n}$$

Corresponding to this finite collection of open sets is a finite collection of points $p_1, p_2, \ldots, p_n$, and a finite collection of positive numbers $\delta_{p_1}, \delta_{p_2}, \ldots, \delta_{p_n}$. Define $\delta = \frac{1}{2} \inf \{\delta_{p_1}, \delta_{p_2}, \ldots, \delta_{p_n}\}$. Let q_1 and q_2 be any two points of A such that $|q_1 - q_2| < \delta$. There is a p_k such that $q_1 \in G_{p_k}$. This means that $|q_1 - p_k| < \delta_{p_k}$. Consequently

$$|q_2 - p_k| \leq |q_2 - q_1| + |q_1 - p_k| < \delta_{p_k} + \delta_{p_k} = 2\delta_{p_k}$$

Therefore

$$|f(q_1) - f(p_k)| < \frac{\epsilon}{2} \qquad \text{and} \qquad |f(q_2) - f(p_k)| < \frac{\epsilon}{2}$$

so that

$$|f(q_1) - f(q_2)| < \epsilon$$

Therefore, f is uniformly continuous on A. QED

We have defined continuous mappings as those mappings which preserve convergent sequences. It is not true, however, that continuous mappings preserve Cauchy sequences, as the student can show by a careful study of Example 3.1. On the other hand, uniformly continuous mappings do preserve Cauchy sequences, and we leave the proof of this fact to the student.

Theorem 3.7. Uniformly continuous mappings preserve Cauchy sequences.

In mathematics, we sometimes have a function f which is defined on the set B of all rational numbers in a closed interval A, and we wish to extend the domain of the function to the set A; that is, we want to find a function F defined on A with the property that $F(x) = f(x)$ for every x in B. Such a function F is called an extension of f from B to A. The problem that we most frequently encounter is that where the given function f is continuous on B, and we wish the extended function F to be continuous on A. We shall find that uniform continuity of f is an essential condition for obtaining the continuous extension F. We now give formal definitions of these concepts.

Definition 3.8. Let A and B be subsets of R_1, $B \subset A$. Suppose that $f: B \to R_1$ is a mapping. Then the mapping $F: A \to R_1$ is said to be an *extension* of f iff $F(x) = f(x)$ for every x in B. If both the functions f and F are continuous, then F is said to be a *continuous extension* of f.

Theorem 3.9. Let $A = [a, b]$, and $B \subset A$. Suppose that $f: B \to R_1$ is a continuous mapping, and that $F: A \to R_1$ is a continuous extension of f. Then both of the mappings f and F are uniformly continuous.

The proof is left as an exercise.

We now proceed at once to our principal theorem.

Theorem 3.10. Let B be the set of all rational numbers in a closed interval $A = [a, b]$, and let $f: B \to R_1$ be a uniformly continuous mapping. Then there is exactly one continuous extension $F: A \to R_1$, and this continuous extension is uniformly continuous.

Proof. We first prove that there is at most one such continuous extension of f. Suppose that F_1 and F_2 are two continuous extensions of f. Let p be an arbitrary point of A. We know that p is the limit of a sequence of rational numbers in A. Thus, let $\langle p_i \rangle$ be a sequence of rational numbers of the closed interval A such that $\lim p_i = p$, and note that $p_i \in B$ for every i. By the continuity of F_1 and F_2, we see that $\lim F_1(p_i) = F_1(p)$ and $\lim F_2(p_i) = F_2(p)$. However, for every i, we have $F_1(p_i) = f(p_i)$ and $F_2(p_i) = f(p_i)$; consequently $F_1(p_i) = F_2(p_i)$. It follows at once that $F_1(p) = F_2(p)$, and since p was an arbitrary point of A, F_1 and F_2 are the same mapping.

To complete the proof of the theorem, we must show that there exists at least one such continuous extension. This means that we must define $F(p)$ for each point p in A. If $p \in B$, we define $F(p) = f(p)$. Suppose $p \notin B$. Let $\langle p_i \rangle$ be a sequence of points of B such that $\lim p_i = p$. We see that $\langle p_i \rangle$ is a Cauchy sequence. Thus since f is uniformly continuous on B, it follows from Theorem 3.7 that $\langle f(p_i) \rangle$ is a Cauchy sequence. By the Cauchy Criterion for Convergence, $\langle f(p_i) \rangle$ has a unique limit z. It is important that we show that the point z is independent of the choice of the sequence $\langle p_i \rangle$. To do this, let $\langle q_i \rangle$ be any other sequence of points of B converging to the point p. We must show that the sequence $\langle f(q_i) \rangle$ also converges to the point z. We do so by defining a new sequence $\langle x_i \rangle$ obtained by "interlacing" the sequences $\langle p_i \rangle$ and $\langle q_i \rangle$; that is,

$$x_{2i-1} = p_i$$
$$x_{2i} = q_i$$

The student should show that the sequence $\langle x_i \rangle$ converges to p. Consequently the sequence $\langle x_i \rangle$ is a Cauchy sequence. Therefore the sequence $\langle f(x_i) \rangle$ is a Cauchy sequence, and we know that the subsequence $\langle f(x_{2i-1}) \rangle$ is the sequence $\langle f(p_i) \rangle$, which converges to z. By Theorem 3.5 of Chapter IV, every subsequence of $\langle f(x_i) \rangle$ converges to z. In particular, $\langle f(x_{2i}) \rangle$ converges to z; that is, $\langle f(q_i) \rangle$ converges to z. Therefore $w = z$, and we have shown that the point z is independent of the choice of the sequence converging to p.

We note that z is a point of R_1, and we define $F(p) = z$. It should be noted that we have defined a mapping $F: A \to R_1$ with the property that $F(x) = f(x)$ for every x in B. Thus F is an extension of f. To complete the proof, we shall show that F is uniformly continuous. To do this, let $\epsilon > 0$ be given. Then there exists a $\delta > 0$ such that if x and x' are any two points of B with $|x - x'| < \delta$, then $|f(x) - f(x')| < \epsilon/4$. This is true, of course, since f is uniformly continuous on B. Let y and y' be any two points of A such that $|y - y'| < \delta/2$. Our proof will be complete if we show that $|F(y) - F(y')| < \epsilon$. Let $\langle p_i \rangle$ be a sequence of points of B converging to the point y, and $\langle q_i \rangle$ a sequence of points of B converging to the point y'. Then $\lim f(p_i) = F(y)$, and $\lim f(q_i) = F(y')$. Therefore there exist positive integers N_1, N_2, N_3, N_4 such that

$$(1) \qquad |f(p_i) - F(y)| < \frac{\epsilon}{4} \qquad \text{for } i > N_1$$

$$(2) \qquad |f(q_i) - F(y')| < \frac{\epsilon}{4} \qquad \text{for } i > N_2$$

$$(3) \qquad |p_i - y| < \frac{\delta}{4} \qquad \text{for } i > N_3$$

$$(4) \qquad |q_i - y'| < \frac{\delta}{4} \qquad \text{for } i > N_4$$

Define $N = \sup \{N_1, N_2, N_3, N_4\}$, and note that if $i > N$, then (1), (2), (3), and (4) all hold. We also recall that $|y - y'| < \delta/2$. Therefore by (3) and (4), we have for each $i > N$

$$|p_i - q_i| \leq |p_i - y| + |y - y'| + |y' - q_i| < \frac{\delta}{4} + \frac{\delta}{2} + \frac{\delta}{4} = \delta$$

Consequently $|f(p_i) - f(q_i)| < \epsilon/4$ for $i > N$. Finally, we see that

$$\begin{aligned} |F(y) - F(y')| &\leq |F(y) - f(p_i)| + |f(p_i) - f(q_i)| + |f(q_i) - F(y')| \\ &< \frac{\epsilon}{4} + \frac{\epsilon}{4} + \frac{\epsilon}{4} \\ &< \epsilon \end{aligned}$$

Therefore F is uniformly continuous. QED

One of the useful applications of this theorem is in the definition of the exponential function a^x for all real values of x, when a is any positive number not equal to 1. We shall confine our attention to the special case $a = 2$. A little reflection on the development of the exponential function in algebra should convince the student that although the definition and properties of 2^x were presented for rational numbers x, no mention was made of numbers such as 2^π or $2^{\sqrt{2}}$. We shall ultimately define a function $F: R_1 \to R_1$ such that $F(x) = 2^x$ for every real number x. As a first step in this direction, we define K as the set of all rational numbers. We shall define a mapping $E: K \to R_1$ such that $E(x) = 2^x$ for $x \in K$. As a preliminary to this definition, we define the mapping E on the subset I_o of K.

Definition 3.11. For each $n \in I_o$, $E(n) = 2 \cdot 2 \cdot 2 \cdot \ldots \cdot 2$, n factors. We also define $2^n = E(n)$ for $n \in I_o$.

The student is familiar with the following facts:

$$2^x 2^y = 2^{x+y} \qquad \text{where } x,y \in I_o$$
$$2^1 = 2$$

Stated in our new notation, these may be written as

$$(5) \qquad E(x)E(y) = E(x+y) \qquad \text{for all } x,y \in I_o$$
$$(6) \qquad E(1) = 2$$

As we extend E to the set K, we shall insist that (5) and (6) remain true for all rational numbers x and y. If (5) is to be true for all rational numbers x,y, it must be true when $x = 1$ and $y = 0$. Placing these values in (5) gives us

$$E(1)E(0) = E(1+0) = E(1)$$

which can be written, by virtue of (6), as

$$2E(0) = 2$$

For this reason, we must have

$$E(0) = 1$$

and this becomes our definition of $E(0)$.

Definition 3.12. $E(0) = 1$.

Lemma 3.13. For $x \in I_o$, $E(x) > 0$.

We next define E on the set of all negative integers.

Definition 3.14. For each positive integer x, we define $E(-x) = \dfrac{1}{E(x)}$.

The proof of the following lemma is left as an exercise.

Lemma 3.15. (*a*) Equations (5) and (6) are true for *all* integers x and y.

(*b*)
$$E(y-x) = \frac{E(y)}{E(x)}$$

where x and y are integers.

(*c*) If x and y are integers $x < y$, then $E(x) < E(y)$.

(*d*) $E(x) > 0$ for every integer x.

We emphasize the fact that the set of all integers is the union of the set of all positive integers, the set of all negative integers, and the single integer zero. It is to be noted that by our definition, for any positive integer n,

$$E(-n) = \frac{1}{E(n)} = \frac{1}{2^n}$$

Definition 3.16. $2^{-n} = E(-n) = \dfrac{1}{2^n}.$

To find the value of $E(1/k)$, where k is a positive integer, we consider the continuous function $f\colon R_1 \to R_1$ defined by $f(x) = x^k$. We note that if x_1 and x_2 are positive numbers and if $x_1 < x_2$, then $f(x_1) < f(x_2)$. In particular, this means that there is at most one positive number x such that $x^k = 2$. To show that there is at least one such x, we note that $f(0) = 1 < 2$, and $f(3) = 2^3 > 2$. Therefore, by the Intermediate Value Theorem, there exists a positive number x_0 such that $x_0^k = 2$. Using this value of x_0, we may now formally define $E(1/k)$ and $2^{1/k}$.

Definition 3.17. (*a*) $E\left(\dfrac{1}{k}\right) = x_0$, where x_0 is the unique positive solution of the equation $x^k = 2$.

(*b*)
$$2^{1/k} = E\left(\frac{1}{k}\right)$$

Lemma 3.18. $E\left(\dfrac{1}{k}\right) > 1$, for every positive integer k.

Definition 3.19. If k is a positive integer, $E\left(-\dfrac{1}{k}\right) = \dfrac{1}{E(k)}.$

It can be shown that for any two positive integers p and q, the equation $x^q = 2^p$ has a unique positive solution. We define $E\left(\dfrac{p}{q}\right)$ to be this solution, and state our definition formally as follows.

Definition 3.20. (*a*) If p and q are positive integers, then $E\left(\dfrac{p}{q}\right)$ is the unique positive solution of the equation $x^q = 2^p$.

(*b*)
$$2^{p/q} = E\left(\frac{p}{q}\right)$$

(c)
$$E\left(-\frac{p}{q}\right) = \frac{1}{E\left(\dfrac{p}{q}\right)}$$

(*d*) $$2^{-(p/q)} = E\left(-\frac{p}{q}\right) = \frac{1}{2^{p/q}}$$

Lemma 3.21. If $p/q = p'/q'$, where p, q, p', q' are positive integers, then $2^{p/q} = 2^{p'/q'}$.

Proof. We define $a_1 = 2^{p/q} = (2^p)^{1/q}$, and note that $a_1^q = 2^p$. From this it follows that

(7) $$a_1^{qq'} = 2^{pq'}$$

Now define $a_2 = 2^{p'/q'} = (2^{p'})^{1/q'}$, and note that $a_2^{q'} = 2^{p'}$. From this it follows that

(8) $$a_2^{qq'} = 2^{p'q}$$

The statement $p/q = p'/q'$ means precisely that $pq' = p'q$. Thus from (7) and (8), we have at once

(9) $$a_1^{qq'} = a_2^{qq'}$$

Let z be the common value of $a_1^{qq'}$ and $a_2^{qq'}$. Then each of the positive numbers a_1 and a_2 is a solution of the equation $x^{qq'} = z$. Since the equation has exactly one positive root, we see that $a_1 = a_2$. QED

Corollary 3.22. If $p/q = p'/q'$, where p, q, p', q' are positive integers, then $E\left(\frac{p}{q}\right) = E\left(\frac{p'}{q'}\right)$.

Lemma 3.23. $\frac{E(y)}{E(x)} = E(y - x)$, that is, $\frac{2^y}{2^x} = 2^{y-x}$, where x and y are rational numbers.

Proof. By virtue of Corollary 3.22, x and y may be written as fractions with the same denominator. Thus, let $x = p/q$, $y = r/q$.

Case 1. $y - x \geq 0$. If $y - x = 0$, the result is trivial. Thus we may assume in this case that $y - x = (r - p)/q > 0$. We must prove that $2^{r/q}/2^{p/q} = 2^{(r-p)/q}$. Let $a_1 = 2^{r/q}$ and $a_2 = 2^{p/q}$. This means that $a_1^q = 2^r$ and $a_2^q = 2^p$. Accordingly, since $r > p$,

$$\left(\frac{a_1}{a_2}\right)^q = \frac{a_1^q}{a_2^q} = \frac{2^r}{2^p} = 2^{r-p}$$

Consequently, a_1/a_2 is the unique positive solution of the equation $x^q = 2^{r-p}$, where $r - p$ is a positive integer. Therefore $a_1/a_2 = 2^{(r-p)/q}$, and this completes the proof of Case 1.

Case 2. $y - x < 0$. In this case, we note that $x - y > 0$, so that by Case 1 we have

$$E(y - x) = \frac{1}{E(x - y)} = \frac{1}{E(x)/E(y)} = \frac{E(y)}{E(x)} \qquad \text{QED}$$

Lemma 3.24. For p and q positive integers, we have

(*a*) $$\left[E\left(\frac{p}{q}\right)\right]^q = 2^p$$

(*b*) $$E\left(\frac{p}{q}\right) = 2^{p/q} = (2^p)^{1/q} > 1$$

We have now defined both $E(x)$ and 2^x in such a way that $2^x = E(x)$ for each rational number x. We need the following important theorem.

Theorem 3.25. If x is any rational number, then $E(x) = 2^x$. Moreover, for any two rational numbers x and y, we have

(*a*) $E(1) = 2$, that is, $2^1 = 2$
(*b*) $E(x) > 0$, that is, $2^x > 0$
(*c*) $E(x)E(y) = E(x + y)$, that is, $2^x 2^y = 2^{x+y}$

Proof. (*a*) is immediate, and the proof of (*b*) is left as an exercise. To prove (*c*), we make use of Lemma 3.23, and note that

$$\begin{aligned} E(x + y) &= E[x - (-y)] \\ &= \frac{E(x)}{E(-y)} \\ &= E(x)E(y) \end{aligned} \qquad \text{QED}$$

Lemma 3.26. If x is any positive rational number, then $E(x) > 1$.

Proof. This is a restatement of Lemma 3.24(*b*).

Lemma 3.27. If x and y are rational numbers with $x < y$, then $E(x) < E(y)$.

Proof. We have at once

$$\begin{aligned} E(y) - E(x) &= E(x)\left[\frac{E(y)}{E(x)} - 1\right] \\ &= E(x)[E(y - x) - 1] \quad \text{by Lemma 3.23} \\ &> 0 \quad \text{by Lemmas 3.25}(b) \text{ and } 3.26 \end{aligned} \qquad \text{QED}$$

Now that we have established all the algebraic properties of E, we are ready to deal with limits.

Lemma 3.28. Each of the sequences $\left\langle E\left(\frac{1}{2^n}\right)\right\rangle$ and $\left\langle E\left(-\frac{1}{2^n}\right)\right\rangle$ converges to 1.

Proof. We see at once by Lemmas 3.26 and 3.27 that the first sequence is a monotone decreasing sequence, every term of which is greater than 1. We recall that every monotone decreasing sequence which has a lower bound converges to its infimum. Therefore the sequence $\langle E(1/2^n)\rangle$ converges to a real number $b \geq 1$. In particular, $b \neq 0$. We may thus write

$$b = \lim E\left(\frac{1}{2^n}\right)$$

Consequently,

$$\begin{aligned} b^2 &= \lim E\left(\frac{1}{2^n}\right) \lim E\left(\frac{1}{2^n}\right) \\ &= \lim \left[E\left(\frac{1}{2^n}\right) E\left(\frac{1}{2^n}\right)\right] \\ &= \lim E\left(\frac{1}{2^n} + \frac{1}{2^n}\right) \\ &= \lim E\left(\frac{1}{2^{n-1}}\right) \\ &= b, \end{aligned}$$

since the sequence $\langle E(1/2^{n-1})\rangle$ is a subsequence of $\langle E(1/2^n)\rangle$. We have shown that $b^2 = b$, and $b \neq 0$. Therefore $b = 1$. It then follows immediately from the fact that $E(-x) = 1/E(x)$ and the quotient theorem for limits that $\langle E(-1/2^n)\rangle$ converges to 1.

QED

Lemma 3.29. The function E defined for all rational numbers is continuous at $x = 0$.

Proof. We know that $E(0) = 1$. Pick any $\epsilon > 0$. We must find a $\delta > 0$ such that if $-\delta < x < \delta$, then $1 - \epsilon < E(x) < 1 + \epsilon$. Since each of the sequences $\langle E(1/2^n)\rangle$ and $\langle E(-1/2^n)\rangle$ converges to 1, we know that there exist integers N_1 and N_2 such that

$$1 - \epsilon < E\left(-\frac{1}{2^{N_1}}\right) < 1 < E\left(\frac{1}{2^{N_2}}\right) < 1 + \epsilon$$

Define $\delta = \inf \{1/2^{N_1}, 1/2^{N_2}\}$. Let x be any rational number such that $-\delta < x < \delta$. Then $-1/2^{N_1} < x < 1/2^{N_2}$. Consequently, by Lemma 3.27,

$$1 - \epsilon < E\left(-\frac{1}{2^{N_1}}\right) < E(x) < E\left(\frac{1}{2^{N_2}}\right) < 1 + \epsilon$$

and we have proven that

$$1 - \epsilon < E(x) < 1 + \epsilon \qquad \text{QED}$$

Theorem 3.30. The function E defined for all rational numbers is uniformly continuous on the set consisting of all rational numbers in the closed interval $[-N, N]$ for every positive integer N.

Proof. Pick any $\epsilon > 0$. There exists a $\delta > 0$ such that if x is any rational number with $-\delta < x < \delta$, then $|1 - E(x)| < \epsilon/2^N$. Let r_1 and r_2 be any two rational numbers in the closed interval $[-N, N]$ and suppose $r_2 < r_1$, and that $|r_1 - r_2| < \delta$. Then

$$\begin{aligned} |E(r_2) - E(r_1)| &= E(r_2)|1 - E(r_1 - r_2)| \\ &< 2^N \cdot \frac{\epsilon}{2^N} \\ &= \epsilon \end{aligned} \qquad \text{QED}$$

Now let N be any positive integer. We have seen that the function E is defined and uniformly continuous on the set of all rational numbers in the closed interval $[-N, N]$. By Theorem 3.10, there exists *exactly one* continuous mapping $F_N: [-N, N] \to R_1$ having the property that $F_N(x) = E(x)$ for every rational number x in this closed interval. Another way of stating this is to observe that $F_N(x)$ is defined and continuous for all real numbers x in the closed interval $[-N, N]$. Furthermore, $F_N(x) = 2^x$ for every rational number in this interval. Moreover, F_N is the only continuous function defined on the closed interval $[-N, N]$ which has the properties (5) and (6) for all rational numbers x and y in this closed interval.

Now consider the sequence of functions $\langle F_1, F_2, F_3, \ldots \rangle$. It should be noted that for each integer n, we have the mapping $F_n: [-n, n] \to R_1$. It should be shown by the student that for any two positive integers h and k, $h < k$, we have $F_k(x) = F_h(x)$ for every real number x in the closed interval $[-h, h]$. In other words, F_k is an extension of F_h.

We now define a mapping $F: R_1 \to R_1$ in the following way.

Given any $x \in R_1$, let n be any integer such that x is in the closed interval $[-n, n]$. Define $F(x) = F_n(x)$.

The properties of F stated in the next theorem are easily established, and are left as exercises.

Theorem 3.31. The function $F: R_1 \to R_1$ just defined has the following properties:

(*a*) $F(x) = 2^x$ for every rational number x.

(*b*) F is continuous.

(*c*) F is the only function of R_1 into R_1 which satisfies both (*a*) and (*b*).

We are now prepared to define formally the exponential function 2^x.

Definition 3.32. For each real number x, we define $2^x = F(x)$, where F is the function defined above.

Theorem 3.33. The continuous mapping $F: R_1 \to R_1$ defined by $F(x) = 2^x$ for every x has the following properties for every pair of real numbers x and y:

(*a*) $F(x + y) = F(x) \cdot F(y)$; that is, $2^{x+y} = 2^x \cdot 2^y$.

(*b*) $F(x) > 0$; that is, $2^x > 0$.

(*c*) If $x > 0$, then $F(x) > 1$.

(*d*) If $x < y$, then $F(x) < F(y)$; that is, $2^x < 2^y$.

Proof. To prove (*a*), let $\langle x_n \rangle$ and $\langle y_i \rangle$ be sequences of rational numbers such that $\lim x_n = x$ and $\lim y_i = y$. Since the values of F and E are the same for all rational numbers, we have for each n and i

$$\begin{aligned} F(x_n + y_i) &= E(x_n + y_i) \\ &= E(x_n) \cdot E(y_i) \qquad \text{by (5)} \\ &= F(x_n) \cdot F(y_i) \end{aligned}$$

We now hold n fixed, and take the limit on i. Since F is continuous, we have for each fixed n

$$\begin{aligned} F(x_n + y) &= \lim_i F(x_n + y_i) \\ &= \lim_i [F(x_n) \cdot F(y_i)] \\ &= F(x_n) \lim_i F(y_i) \\ &= F(x_n) \cdot F(y) \end{aligned}$$

We next take the limit on n, again using the continuity of F.

$$\begin{aligned} F(x+y) &= \lim_n F(x_n+y) \\ &= \lim_n [F(x_n)\cdot F(y)] \\ &= F(x)\cdot F(y) \end{aligned}$$

This completes the proof of (*a*).

The proof of (*b*) follows easily from (*a*), and is left as an exercise.

To prove (*c*), let $\langle r_i \rangle$ be any sequence of positive rational numbers such that $\lim r_i = x > 0$. We note by Lemma 3.24(*b*) that $F(r_i) > 1$ for every i. Therefore $\lim F(r_i) = F(x) \geq 1$. It remains to be shown that $F(x) \neq 1$. By the Archimedean property, since $x > 0$, there exists a positive integer k such that $kx > 2$. Thus, assuming $F(x) = 1$, we have

$$F(kx) = [F(x)]^k = 1 \tag{10}$$

Since $kx > 2$, there exists a sequence of rational numbers $\langle p_i \rangle$ such that $\lim p_i = kx$, and $p_i > 2$ for every i. It follows from Lemma 3.27 that $F(p_i) > F(2) = 4$ for every i. Using (10), we obtain the contradiction $1 = F(kx) = \lim F(p_i) \geq 4$. Thus the proof of (*c*) is complete.

The proof of (*d*) is left as an exercise. QED

We have now shown how Theorem 3.10 may be used to define the exponential function 2^x for all real numbers x. An essentially identical proof may be used to define the exponential function a^x, where a is any real number greater than 1, the principal change consisting of replacing (6) by $E(1) = a$. To define a^x when $0 < a < 1$ is quite similar; however, some of the inequalities are reversed.

PROBLEMS

44. (*a*) Let A and f be defined as in Example 3.1, and let $\epsilon = 1/100$. Find a suitable δ for each of the following points p:

(1) $p = \frac{1}{2}$ (2) $p = \frac{1}{3}$ (3) $p = \frac{1}{4}$ (4) $p = \frac{1}{10}$ (5) $p = \frac{1}{10{,}000}$.

(*b*) Find a δ which will work for all the values of p listed above.

(*c*) Given $\epsilon = 10$, $\delta = 1/1000$, find a point p in A and a point x in A such that $|x - p| < \delta$ but $|f(x) - f(p)| > \epsilon$.

(*d*) Given any $\epsilon > 0$ and any $\delta > 0$, show that there exists a point p in A and a point x in A such that $|p - x| < \delta$ but $|f(p) - f(x)| > \epsilon$. Conclude that f is not uniformly continuous.

45. Let $A = (0, 1)$, and $g: A \to R_1$ be defined by $g(x) = 1/(1 - x)$ for each x in A. Given a point $p \in A$ and a real number $\epsilon > 0$, find an explicit formula for $\delta > 0$ such that if $|x - p| < \delta$, then $|f(x) - f(p)| < \epsilon$.

46. Let A be any uniformly isolated sequence of points, and let $f: A \to R_1$ be any mapping. Prove that f is continuous by using Theorem 2.19. Then show that the δ which you find depends neither on the $\epsilon > 0$ which you use nor on the point p at which you establish the continuity of the mapping. Thus conclude that f is uniformly continuous.

47. State exactly what is meant by the statement: "The mapping $f: A \to R_1$ is not uniformly continuous."

48. Prove Theorem 3.3.

49. Give an example of a continuous mapping $f: A \to R_1$ and of a Cauchy sequence $\langle a_i \rangle$ in A such that $\langle f(a_i) \rangle$ is not a Cauchy sequence.

50. Prove Theorem 3.7.

51. Prove Theorem 3.9.

52. In the proof of Theorem 3.10, show that the sequence $\langle x_i \rangle$ converges to p.

53. Prove Lemma 3.13.

54. Prove Lemma 3.15.

55. Show that if $f: R_1 \to R_1$ is the mapping defined by $f(x) = x^k$, where k is a positive integer, and if $x_1 < x_2$, then $f(x_1) < f(x_2)$. (*Hint:* Factor the expression $x_2^k - x_1^k$.)

56. Prove that for any two positive integers p and q, the equation $x^q = 2^p$ has a unique positive solution.

57. Prove Lemma 3.18.

58. In the discussion following Theorem 3.25, show that for the sequence of functions $\langle F_i \rangle$ and for any two integers h and k, $h < k$, F_k is an extension of F_h.

59. Prove Lemma 3.24.

60. Prove Theorem 3.25(*b*).

61. Prove Theorem 3.31.

62. Prove Theorem 3.33(*b*).

63. Prove Theorem 3.33(*d*).

64. Make a study of the continuous function f defined on all rational numbers and satisfying the condition $f(x + y) = f(x) + f(y), f(1) = a \neq 0$. Find its value at each integer, and then at each rational. What would be the general appearance of the extension of this function to the set of all real numbers?

4. *An application of continuous mappings*

Throughout this section, J denotes the closed interval $[0, 1]$, and $f: J \to J$ a continuous mapping having the following property:

(a) Given any $x \in J$, there exists a positive integer n depending on x such that $f^n(x) = x$.

Here, f^n is the composite function. Thus, for any $x \in J$,

$$f^2(x) = f[f(x)], \qquad f^3(x) = f[f^2(x)], \text{ etc.}$$

The purpose of this section is to determine additional properties of f which follow from those listed. The derivation of these properties involves a review of many of the concepts and techniques which we have developed.

Lemma 4.1. The function f is one-to-one; that is, for any two points x_1 and x_2 of J, $x_1 \neq x_2$, we must have $f(x_1) \neq f(x_2)$.

Proof. Suppose f is not one-to-one. Then there exist points x_1, x_2 in J such that $x_1 \neq x_2$, but $f(x_1) = f(x_2)$. By property (a), there exist positive integers n_1 and n_2 such that

$$f^{n_1}(x_1) = x_1 \qquad \text{and} \qquad f^{n_2}(x_2) = x_2$$

We note that

$$\begin{aligned} f^{2n_1}(x_1) &= f^{n_1}[f^{n_1}(x_1)] = x_1 \\ f^{3n_1}(x_1) &= f^{n_1}[f^{2n_1}(x_1)] = x_1 \\ &\vdots \\ f^{kn_1}(x_1) &= f^{n_1}[f^{(k-1)n_1}(x_1)] = x_1 \end{aligned}$$

In particular, this tells us that $f^{n_2n_1}(x_1) = x_1$. In a similar manner, $f^{n_2n_1}(x_2) = x_2$. Since $f(x_1) = f(x_2)$, however, then $f^{n_2n_1}(x_1) = f^{n_2n_1}(x_2)$, that is, $x_1 = x_2$. This contradiction completes the proof of the lemma.

QED

Lemma 4.2. f is a mapping of J onto J.

The proof is left as an exercise.

Lemma 4.3. The function f is monotone; that is,

(i) $\qquad f(x_1) < f(x_2) \quad$ for every $x_1 < x_2$

or

(ii) $\qquad f(x_1) > f(x_2) \quad$ for every $x_1 < x_2$

Proof. Suppose the lemma is false. Then there exist two points $a, b \in J$ such that

$$a < b \quad \text{and} \quad f(a) - f(b) < 0$$

and there exist two points $c, d \in J$ such that

$$c < d \quad \text{and} \quad f(c) - f(d) > 0$$

Now define the function $h: J \to R_1$ by

$$h(t) = f[a + t(c - a)] - f[b + t(d - b)]$$

[The student should show that $a + t(c - a)$ and $b + t(d - b)$ are elements of J for every $t \in J$.] Note that h is continuous; moreover,

$$h(0) = f(a) - f(b) < 0$$

and

$$h(1) = f(c) - f(d) > 0$$

By the Intermediate Value Theorem, there exists an element $t_0 \in J$ such that $h(t_0) = 0$, that is,

$$f[a + t_0(c - a)] = f[b + t_0(d - b)]$$

Since f is one-to-one, we see at once that

$$a + t_0(c - a) = b + t_0(d - b)$$

This equation may be written in the form

$$-t_0(a + d) = (b - a)(1 - t_0)$$

This is a contradiction, since the left side of this equation is a negative number, whereas the right side is a nonnegative. QED

We say that f is an *increasing function* on J iff condition (i) of Lemma 4.3 holds, and that f is a *decreasing function* on J iff condition (ii) of Lemma 4.3 holds. Lemma 4.3 thus tells us that the function f we are considering must be either an increasing function on J or a decreasing function on J.

Theorem 4.4. If f is any continuous increasing function on J satisfying (a), then f is the identity mapping; that is, $f(x) = x$ for every x in J.

Proof. Suppose there exists a point $x_1 \in J$ such that $f(x_1) \neq x_1$. We must have either $f(x_1) > x_1$ or $f(x_1) < x_1$. We shall discuss the first case, and leave the second as an exercise. Thus suppose that

$x_1 < f(x_1)$. Since f is an increasing function, we must have $f(x_1) < f[f(x_1)] = f^2(x_1)$, so that

$$x_1 < f(x_1) < f^2(x_1)$$

By induction, for each $n \in I_o$,

$$x_1 < f(x_1) < f^2(x_1) < \ldots < f^n(x_1)$$

Consequently, $f^n(x_1) \neq x_1$ for any positive integer n. This contradicts the definition of the mapping f. QED

Theorem 4.5. If f is any continuous decreasing function on J satisfying (a), then f^2 is the identity mapping; that is, $f^2(x) = x$ for every x in J.

Proof. This is immediate from Theorem 4.4, since f^2 is an increasing function on J which satisfies all the conditions which we placed on our original function f. QED

We have the following corollaries to Theorems 4.4 and 4.5.

Corollary 4.6. The positive integer n defined in condition (a) can never be greater than 2.

Corollary 4.7. For every continuous function f satisfying condition (a), f^2 is the identity mapping.

From Lemmas 4.1 and 4.2, we know that for every $y \in J$, there is exactly one x in J such that $f(x) = y$. We now define the mapping $g: J \to J$ by $g(y) = x$. Another way of stating this is to define $g(y) = f^{-1}[\{y\}]$ for every $y \in J$. The mapping g is frequently written as $g = f^{-1}$, and we refer to this mapping as the *inverse mapping* of f.

It should be evident that for any function f of the type we are considering, we must have

$$f^{-1}[f(x)] = f[f^{-1}(x)] = x \text{ for every } x \in J$$

We also have, from Corollary 4.7, the following result.

Corollary 4.8. For every continuous function f satisfying condition (a), we must have $f^{-1} = f$.

We recall that the graph of a function f consists of all points $(x, f(x))$ in the plane of analytical geometry. The graph of f^{-1} thus consists of all points $(f(x), x)$ in the same plane. It follows at once that the graph of f^{-1} is the reflection of the graph of f in the line whose equation is $y = x$. In the case under discussion, $f^{-1} = f$, so that we have the following theorem.

Theorem 4.9. For every continuous function f satisfying condition (a), the graph of f is symmetric with respect to the line whose equation is $y = x$.

We now give a summarizing discussion of those continuous functions f satisfying condition (a). The case in which f is the identity mapping is trivial, and the equation of the graph is then given by $y = x$, $0 \leq x \leq 1$. The case in which f is a decreasing function on J is more interesting. How do we characterize those continuous, decreasing functions defined on J which have the property that f^2 is the identity mapping, that is, that $f^2(x) = x$ for every x in J? This equation tells us that 0 and 1 are in the range of f, and since f (being decreasing) takes its maximum value at 0 and its minimum value at 1, we see that $f(x)$ decreases from 1 to 0 as x increases from 0 to 1. Thus the graph of f is a continuous arc that decreases from the point (0, 1) until it intersects the line $y = x$. The remainder of the graph is obtained by reflecting this arc about the line $y = x$. (See Figure 4.)

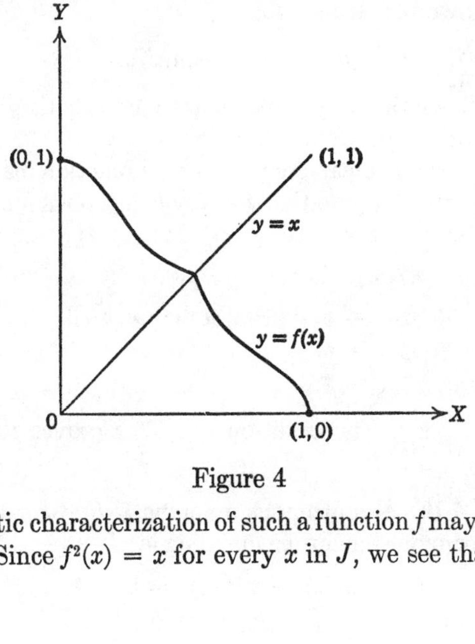

Figure 4

An analytic characterization of such a function f may be obtained as follows. Since $f^2(x) = x$ for every x in J, we see that the equations

(*b*) $$y = f(x)$$
(*c*) $$f(y) = x$$

are equivalent in the sense that they have the same graph. If we subtract corresponding sides of equation (*b*) from equation (*c*) and rearrange, we obtain the equation

(*d*) $$[f(x) - x] + [f(y) - y] = 0$$

Let us write $m(x) = f(x) - x$, and note that equation (*d*) can be written as

(*e*) $$m(x) + m(y) = 0$$

We leave it as an exercise to show that the function m just defined is continuous, decreasing, and maps the closed interval $[0, 1]$ onto the closed interval $[-1, 1]$. We saw that if (x, y) satisfies (*b*), then it also satisfies (*e*); in other words, if x is any point of the closed interval $[0, 1]$, then the following equation holds.

(*f*) $$m(x) + m[f(x)] = 0$$

Since m is decreasing, it is easily seen that m is one-to-one; consequently m has an inverse function, which we denote by m^{-1}. Using equation (*f*), we see at once that

(*g*) $$f(x) = m^{-1}[-m(x)]$$

We have shown that any continuous function satisfying (*a*) must satisfy (*f*), where m is defined above.

Now let m be any continuous, monotone function mapping $[0, 1]$ onto $[-1, 1]$, and suppose that f is any continuous mapping satisfying (*f*). We see at once, using (*g*), that

$$\begin{aligned} f^2(x) &= m^{-1}\{-m[f(x)]\} \\ &= m^{-1}\{-m[m^{-1}(-m(x))]\} \\ &= m^{-1}[m(x)] \\ &= x \end{aligned}$$

Consequently f satisfies condition (*a*). This proves the following theorem.

Theorem 4.10. A continuous mapping $f: J \rightarrow J$ satisfies (*a*) iff there is a continuous monotone mapping $m: J \rightarrow [-1, 1]$ such that

$$m(x) + m[f(x)] = 0$$

Example 4.11. Define $m(x) = \cos x$. Then, by (g), $f(x) = \text{Arccos}\,(-\cos x) = \text{Arccos}\,[\cos\,(\pi - x)] = \pi - x$, since $0 \leq x \leq 1$. Accordingly,

$$f^2(x) = f[f(x)] = f(\pi - x) = \pi - (\pi - x) = x$$

PROBLEMS

65. Prove Lemma 4.2.

66. In the proof of Lemma 4.3, show that $a + t(c - a)$ and $b + t(d - b)$ are elements of J for every $t \in J$.

67. In the proof of Lemma 4.3, prove that the mapping $h: J \rightarrow R_1$ is continuous.

68. Complete the proof of Theorem 4.4 for the case where $f(x_1) < x_1$.

69. In equation (e), prove that the function m is continuous, decreasing, and maps the closed interval $[0, 1]$ onto the closed interval $[-1, 1]$.

70. Using a method similar to that used in Example 4.11, define $m(x) = \frac{1}{2} - x^r$, where r is a positive integer. Find $f(x)$ and prove that $f^2(x) = x$.

71. Prove that if $f: J \rightarrow J$ is a continuous mapping satisfying condition (a), then $f^2: J \rightarrow J$ is a continuous mapping satisfying condition (a).

72. Use Problem 23 to prove that if f is a decreasing continuous function satisfying (a), then there is exactly one point $x_0 \in J$ such that $f(x_0) = x_0$.

VI
Metric spaces

1. Cartesian products, metrics, and metric sets

As a preliminary to our discussion of metric spaces, we concern ourselves in this section with the study of metric sets. We shall see shortly that a metric set is merely a set together with a distance function (called a metric) which is defined on pairs of points in the set, and which satisfies certain properties. The value of the function is a real number which is defined to be the distance between the points.

Before formalizing the definition of a metric set, we need some additional terminology and notation.

Definition 1.1. Let U be the universe, and let A and B be sets in U. Then, the *Cartesian product* of the sets A and B (written $A \times B$ and stated as "A cross B") is the set of all ordered pairs defined as follows:

$$A \times B = \{(a, b): a \in A, b \in B\}$$

It is unfortunate that we use the same notation for the ordered pair (a, b) as for the open interval (a, b). However, this ambiguity of notation should present no difficulty, since we shall be careful to distinguish between the two possible interpretations, if it is not clear from the context. We now give some examples of Cartesian products.

Example 1.2. Let $U = \{1, 2, 3, 4, 5, 6\}$, and let A and B be the following sets in U: $A = \{1, 2, 3, 5\}$; $B = \{4, 5, 6\}$. Then, $A \times B$ is the set consisting of the following ordered pairs:

$$\begin{array}{cccc} (1, 4) & (2, 4) & (3, 4) & (5, 4) \\ (1, 5) & (2, 5) & (3, 5) & (5, 5) \\ (1, 6) & (2, 6) & (3, 6) & (5, 6) \end{array}$$

On the other hand, the set $B \times A$ consists of the following ordered pairs:

$$\begin{array}{ccc} (4, 1) & (5, 1) & (6, 1) \\ (4, 2) & (5, 2) & (6, 2) \\ (4, 3) & (5, 3) & (6, 3) \\ (4, 5) & (5, 5) & (6, 5) \end{array}$$

Note that the sets $A \times B$ and $B \times A$ are generally not equal.

Example 1.3. Define U as in Example 1.2, and consider the Cartesian product of U with itself. It is easy to see that $U \times U$ is the following set of 36 ordered pairs.

$$\begin{array}{cccccc} (1, 1) & (1, 2) & (1, 3) & (1, 4) & (1, 5) & (1, 6) \\ (2, 1) & (2, 2) & (2, 3) & (2, 4) & (2, 5) & (2, 6) \\ (3, 1) & (3, 2) & (3, 3) & (3, 4) & (3, 5) & (3, 6) \\ (4, 1) & (4, 2) & (4, 3) & (4, 4) & (4, 5) & (4, 6) \\ (5, 1) & (5, 2) & (5, 3) & (5, 4) & (5, 5) & (5, 6) \\ (6, 1) & (6, 2) & (6, 3) & (6, 4) & (6, 5) & (6, 6) \end{array}$$

Note in particular that the sets $A \times B$ and $B \times A$ of Example 1.2 are subsets of $U \times U$.

Example 1.4. Let the universe be R_1, and consider the Cartesian product of R_1 with itself. We then see that $R_1 \times R_1$ is the set of all ordered pairs of real numbers. Geometrically, if we "cross" the real line with itself, we get the Euclidean plane. Taking the copies of R_1 as coordinate axes, and their point of intersection as the origin, we see that every point in the plane corresponds to an ordered pair of real numbers, and vice versa. This is precisely how we coordinatized the plane in analytic geometry. In fact, we generally refer to the numbers x and y as the Cartesian coordinates of the point (x, y) in $R_1 \times R_1$. The word "Cartesian" is derived from the name of the great French mathematician, René Descartes (1596–1650), who is called the father of analytic geometry.

Definition 1.5. We use the symbol R_2 to denote the Cartesian product of R_1 with itself; that is, $R_2 = R_1 \times R_1$.

For a given universe U, we have seen that the Cartesian product $U \times U$ is a set, so that we may define a mapping (or function) from $U \times U$ into R_1 merely by associating a unique real number with each element of $U \times U$; that is, with each ordered pair of points of U. Such a mapping f may be indicated by $f: U \times U \to R_1$. Now, if x and y are any points of U, we may consider the ordered pair (x, y) as an element of $U \times U$, so that the mapping f is actually defined on pairs of points of U. We are now ready to define a metric on a set U.

Definition 1.6. Let U be the universe. A *metric* ρ on U is a real-valued function (or mapping) $\rho: U \times U \to R_1$ which satisfies the following conditions: For any points x, y, z in U,

(1) $0 \leq \rho(x, y)$
(2) $\rho(x, y) = \rho(y, x)$
(3) $\rho(x, y) = 0$ iff $x = y$
(4) $\rho(x, y) \leq \rho(x, z) + \rho(z, y)$ (Triangle inequality)

The number $\rho(x, y)$ is called the *distance* between the points x and y.

We see from Definition 1.6 that a metric is merely a distance function which measures the undirected distance between each pair of points in the universe. If we read $\rho(x, y)$ as "the undirected distance from x to y," conditions (1) through (4) exhibit those properties that we would normally expect from any measuring device.

Definition 1.7. A *metric set* is a set together with a metric on the set.

Now that we have defined a metric set, one might question the existence of any such sets. It happens that we have been dealing with one throughout all our previous work.

Theorem 1.8. R_1 is a metric set, with the metric $\rho: R_1 \times R_1 \to R_1$ defined by $\rho(x, y) = |x - y|$ for every pair of points x, y in R_1.

We leave it as an exercise for the student to prove that the function ρ just defined is a metric on R_1; that is, to show that ρ satisfies conditions (1) through (4) of Definition 1.6.

Before giving examples of additional metric sets, we prove an inequality which is fundamental for analysis.

Theorem 1.9 (Cauchy-Schwarz Inequality). Let $a_1, a_2, \ldots, a_n$ and $b_1, b_2, \ldots, b_n$ be real numbers for any positive integer n. Then

$$(a_1b_1 + a_2b_2 + \ldots + a_nb_n)^2 \leq (a_1^2 + a_2^2 + \ldots + a_n^2)(b_1^2 + b_2^2 + \ldots + b_n^2)$$

Proof. Let $A = a_1^2 + a_2^2 + \cdots + a_n^2$, $B = b_1^2 + b_2^2 + \cdots + b_n^2$, and $C = a_1b_1 + a_2b_2 + \ldots + a_nb_n$. We are going to rely on our knowledge of quadratic functions of one variable, so we introduce the real variable x and consider expressions of the form $(xa_i + b_i)$ for $i = 1, 2, \ldots, n$. Note that

$$(1) \quad (xa_1 + b_1)^2 + (xa_2 + b_2)^2 + \ldots + (xa_n + b_n)^2 \geq 0$$

and that if the terms on the left are expanded and combined, then (1) may be written as

$$(2) \qquad Ax^2 + 2Cx + B \geq 0$$

If for each $i = 1, 2, \ldots, n$, we have $a_i = 0$, then the Cauchy-Schwarz inequality is trivially true, since each side is 0. Thus we may assume that at least one a_i is nonzero. It then follows that $A > 0$, and hence the left side of (2) is a quadratic function of x. If we designate this function by y, then

$$(3) \qquad y = Ax^2 + 2Cx + B$$

The graph of (3) is a parabola with vertical axis, and, since $A > 0$, the parabola opens upward. Also, $y \geq 0$ by (2), which means that no point of the graph lies below the X-axis. Thus the vertex of the parabola, which represents the minimum value of y, lies either on the X-axis or above the X-axis. In the former case, the parabola is tangent to the X-axis at the vertex, and hence the abscissa of the vertex is a double root of the equation $y = 0$. In the latter case, the equation $y = 0$ has no real roots. Thus in either case, the quadratic equation $y = 0$ does not have two distinct real roots. Let us recall that if f is a quadratic function of the form $f(x) = ax^2 + bx + c$, then the discriminant of the function is denoted by $b^2 - 4ac$. Furthermore, the quadratic equation $f(x) = 0$ has two distinct real roots, or two equal real roots, or no real roots iff the discriminant is positive, zero, or negative, respectively.

Returning to the problem at hand, we see that the discriminant of the function y defined by (3) is $4C^2 - 4AB$, and since the equation $y = 0$ cannot have two distinct real roots, it follows that

$$4C^2 - 4AB \leq 0$$

which reduces to

$$C^2 \leq AB$$

From the definitions of A, B, and C, we see that this is the desired result. QED

Theorem 1.10. The Cartesian product of two metric sets is a metric set.

Proof. Let S_1 and S_2 be metric sets with metrics ρ_1 and ρ_2, respectively, and let $S = S_1 \times S_2$. We thus have $\rho_1: S_1 \times S_1 \to R_1$ and $\rho_2: S_2 \times S_2 \to R_1$. Let p_1, p_2, and p_3 be any three points in S. Then there are points x_1, y_1, z_1 in S_1 and x_2, y_2, z_2 in S_2 such that $p_1 = (x_1, x_2)$, $p_2 = (y_1, y_2)$, and $p_3 = (z_1, z_2)$. Now define $\rho: S \times S \to R_1$ as follows:

$$\rho(p_1, p_2) = \{[\rho_1(x_1, y_1)]^2 + [\rho_2(x_2, y_2)]^2\}^{1/2}$$

We want to show that ρ is a metric on S. The first three conditions of Definition 1.6 follow fairly easily from the fact that ρ_1 and ρ_2 are metrics, so the proofs of these are left as an exercise. We now prove the triangle inequality. Thus we must show that

$$\rho(p_1, p_2) \leq \rho(p_1, p_3) + \rho(p_3, p_2)$$

We first note the following special case of the Cauchy-Schwarz inequality, with $n = 2$:

$$\begin{aligned}
&\rho(p_1, p_3) \cdot \rho(p_3, p_2) \\
&\quad = \{[\rho_1(x_1, z_1)]^2 + [\rho_2(x_2, z_2)]^2\}^{1/2}\{[\rho_1(z_1, y_1)]^2 \\
&\qquad\qquad + [\rho_2(z_2, y_2)]^2\}^{1/2} \\
&\quad \geq \rho_1(x_1, z_1) \cdot \rho_1(z_1, y_1) + \rho_2(x_2, z_2) \cdot \rho_2(z_2, y_2)
\end{aligned}$$

Now using the definition of ρ, the above inequality, and the fact that ρ_1 and ρ_2 are metrics, we have

$$\begin{aligned}
&[\rho(p_1, p_3) + \rho(p_3, p_2)]^2 \\
&\quad = [\rho(p_1, p_3)]^2 + 2\rho(p_1, p_3)\rho(p_3, p_2) + [\rho(p_3, p_2)]^2 \\
&\quad = [\rho_1(x_1, z_1)]^2 + [\rho_2(x_2, z_2)]^2 + 2\rho(p_1, p_3)\rho(p_3, p_2) \\
&\qquad\qquad + [\rho_1(z_1, y_1)]^2 + [\rho_2(z_2, y_2)]^2 \\
&\quad \geq [\rho_1(x_1, z_1)]^2 + [\rho_2(x_2, z_2)]^2 + 2\rho_1(x_1, z_1)\rho_1(z_1, y_1) \\
&\qquad\qquad + 2\rho_2(x_2, z_2)\rho_2(z_2, y_2) + [\rho_1(z_1, y_1)]^2 + [\rho_2(z_2, y_2)]^2
\end{aligned}$$

$$
\begin{aligned}
&= [\rho_1(x_1, z_1) + \rho_1(z_1, y_1)]^2 + [\rho_2(x_2, z_2) + \rho_2(z_2, y_2)]^2 \\
&\geq [\rho_1(x_1, y_1)]^2 + [\rho_2(x_2, y_2)]^2 \\
&= [\rho(p_1, p_2)]^2
\end{aligned}
$$

Taking square roots and recalling Definition 1.6 (1), we thus have

$$\rho(p_1, p_2) \leq \rho(p_1, p_3) + \rho(p_3, p_2) \qquad \text{QED}$$

Let us take particular notice of R_2 for a moment. If p_1 and p_2 are any two points of R_2, then let $p_1 = (x_1, y_1)$ and $p_2 = (x_2, y_2)$. Denoting ρ_1 as the metric on R_1, where $\rho_1: R_1 \times R_1 \to R_1$, we see that

$$\rho_1(x_1, x_2) = |x_1 - x_2| \qquad \text{and} \qquad \rho_1(y_1, y_2) = |y_1 - y_2|$$

so that $[\rho_1(x_1, x_2)]^2 = (x_1 - x_2)^2$ and $[\rho_1(y_1, y_2)]^2 = (y_1 - y_2)^2$. Hence the metric on R_2 as defined in Theorem 1.10 is

$$\rho(p_1, p_2) = [(x_1 - x_2)^2 + (y_1 - y_2)^2]^{1/2}$$

Note that this is precisely the formula for the undirected distance between the points p_1 and p_2 as developed in analytic geometry. We thus have the following corollary to Theorem 1.10.

Corollary 1.11. R_2 is a metric set with the metric described above.

PROBLEMS

1. Prove Theorem 1.8.
2. Prove that the function ρ defined in Theorem 1.10 satisfies the first three conditions of Definition 1.6.
3. Show that any set may be made into a metric set.
4. Let $\rho_1: R_1 \times R_1 \to R_1$ be defined by $\rho_1(x_1, x_2) = |x_1 - x_2|$ for any pair of points x_1, x_2 in R_1. Let us now define the function $\rho: R_2 \times R_2 \to R_1$ as follows:

If $p_1 = (x_1, y_1)$ and $p_2 = (x_2, y_2)$ are any two points in R_2, then

$$\rho(p_1, p_2) = \rho_1(x_1, x_2) + \rho_1(y_1, y_2)$$

(*a*) Prove that ρ is a metric on R_2.
(*b*) Determine all points p in R_2 such that $\rho[p, (0, 0)] \leq 1$.
(*c*) Determine all points p in R_2 such that $\rho[p, (0, 0)] = 1$.

2. *Open and closed sets. Metric spaces*

Throughout the previous five chapters, we restricted our attention almost exclusively to the universe R_1, considering various properties of sets and sequences in R_1, as well as properties of functions defined from R_1 into itself. Upon closer inspection, most of these properties will be seen to depend, either directly or indirectly, on the notion of distance—or, using the terminology of the preceding section, on the fact that R_1 is a metric set.

It is quite natural, then, to expect that many of these same properties may be defined for arbitrary metric sets. In view of the important role of open sets in the development of our theory for R_1, it seems reasonable to begin our discussion of metric sets by defining the concept of an open set in a given metric set. To this end, we first define spherical neighborhoods.

Definition 2.1. Let U be a metric set with metric ρ. For any point p of U and any positive number r, the *spherical neighborhood* $S(p, r)$ consists of all points x of U such that $\rho(p, x) < r$; that is,

$$S(p, r) = \{x \in U: \rho(p, x) < r\}$$

We now give two examples illustrating spherical neighborhoods in the familiar metric sets R_1 and R_2.

Example 2.2. Let $U = R_1$, with the metric ρ defined by $\rho(x, y) = |x - y|$ for x, y in R_1. Then for any $p \in R_1$ and any $r > 0$

$$S(p, r) = \{x \in R_1: |x - p| < r\}$$

or

$$S(p, r) = \{x \in R_1: p - r < x < p + r\}$$

or

$$S(p, r) = \{x \in R_1: x \in (p - r, p + r)\}$$

that is, the spherical neighborhood $S(p, r)$ in R_1 is merely the open interval of length $2r$ and centered at p.

Example 2.3. Let $U = R_2$, with metric ρ defined as follows: Given $p_1 = (x_1, y_1)$ and $p_2 = (x_2, y_2)$ any two points of R_2, $\rho(p_1, p_2) = [(x_1 - x_2)^2 + (y_1 - y_2)^2]^{1/2}$. Then for any point $p_0 = (x_0, y_0)$ in R_2 and any $r > 0$,

$$S(p_0, r) = \{p \in R_2: \rho(p, p_0) < r\}$$

or

$$S(p_0, r) = \{(x, y) \in R_2: [(x - x_0)^2 + (y - y_0)^2]^{1/2} < r\}$$

that is, the spherical neighborhood $S(p_0, r)$ in R_2 is merely the interior of the circle of radius r and center p_0.

Lemma 2.4. Let U be a metric set with metric ρ, let $S(p, r)$ be a spherical neighborhood in U, and let q be any point of $S(p, r)$. Then there exists a real number $r' > 0$ such that $S(q, r') \subset S(p, r)$.

Proof. The fact that $q \in S(p, r)$ implies that $\rho(p, q) < r$. Define $r' = r - \rho(p, q)$. Then $r' > 0$. It remains to show that $S(q, r') \subset S(p, r)$. Thus let x be any point in $S(q, r')$. This means that $\rho(q, x) < r'$. From the definition of r', we see that $\rho(p, q) = r - r'$. Hence, by the triangle inequality, we have

$$\begin{aligned} \rho(p, x) &\leq \rho(p, q) + \rho(q, x) \\ &= r - r' + \rho(q, x) \\ &< r - r' + r' \\ &= r \end{aligned}$$

Since $\rho(p, x) < r$, it follows that $x \in S(p, r)$. Therefore $S(q, r') \subset S(p, r)$. QED

We are now ready to define an open set in a metric set U.

Definition 2.5. A set G in a metric set U is *open* iff given any point $p \in G$, there exists $\epsilon > 0$ such that $S(p, \epsilon) \subset G$.

Given any point p in a metric set U and any $\epsilon > 0$, we know that $S(p, \epsilon) \subset U$. Thus we have the following theorem.

Theorem 2.6. If the universe U is a metric set, then U is open.

Note that Definition 2.5 is consistent with our definition of an open set in R_1, since in R_1 the spherical neighborhood $S(p, \epsilon)$ is merely the open interval $(p - \epsilon, p + \epsilon)$. Just as in R_1 we saw that an open interval is an open set, we now have a corresponding result in an arbitrary metric set.

Theorem 2.7. A spherical neighborhood $S(p, r)$ in a metric set U is an open set.

Proof. The result follows immediately from Lemma 2.4.

Since we shall find it convenient to use the negation of Definition 2.5, we now state it as a theorem.

Theorem 2.8. The set G in the metric set U is not open iff there exists a point $p \in G$ such that given any $\epsilon > 0$, $S(p, \epsilon)$ is not contained in G.

By virtue of Theorem 2.8, we immediately have the following.

Theorem 2.9. The empty set ϕ is open in any metric set U.

Up to this point, we have not considered either the union or the intersection of a collection of sets where the index set is empty.

We now find it convenient to define these concepts in order to avoid exceptional cases in the theorems which are to follow.

Definition 2.10. Let H be an index set, U the universe, and $\{A_\alpha\}$ a collection of subsets of U indexed by H. If $H = \phi$, then $\{A_\alpha\}$ is called the *empty collection* of subsets of U, and we define

$$\bigcup_{\alpha \in H} A_\alpha = \phi, \qquad \bigcap_{\alpha \in H} A_\alpha = U$$

The student should show that DeMorgan's Theorem is true under these generalizations of the definitions of union and intersection.

Theorem 2.11. The union of any collection of open sets in the metric set U is an open set.

Proof. Let H be an index set. If $H = \phi$, the theorem is a corollary of Theorem 2.9. If $H \neq \phi$, then for every $p \in H$, let A_p be an open set in U. Let q be any point in $\bigcup_{p \in H} A_p$. Then q must belong to at least one of the sets A_p; that is, there exists $p_0 \in H$ such that $q \in A_{p_0}$. Since A_{p_0} is open in U by hypothesis, there exists $\epsilon > 0$ such that $S(q, \epsilon) \subset A_{p_0}$. Hence $S(q, \epsilon) \subset \bigcup_{p \in H} A_p$. Therefore $\bigcup_{p \in H} A_p$ is an open set in U, by Definition 2.5. QED

Theorem 2.12. The intersection of any finite collection of open sets in the metric set U is an open set.

Proof. Left as an exercise. Do not overlook the case where the index set is empty.

The fact that the word "finite" is essential in the statement of Theorem 2.12 has already been established for the metric set R_1.

As in the case of R_1, we now define closed sets in terms of open sets.

Definition 2.13. A set E in the metric set U is *closed* iff its complement $U - E$ is open.

By virtue of Theorems 2.6 and 2.9 and the preceding definition, the following result is immediate.

Theorem 2.14. If the universe U is a metric set, then each of the sets U and ϕ is both open and closed.

The following theorems regarding closed sets are easily proved by using DeMorgan's Theorem and the corresponding result for open sets. We thus leave their proofs as exercises.

Theorem 2.15. The intersection of any collection of closed sets in the metric set U is a closed set.

Theorem 2.16. The union of any finite collection of closed sets in the metric set U is a closed set.

The fact that the word "finite" is essential in the statement of Theorem 2.16 has already been established for the metric set R_1.

We might point out that although every spherical neighborhood in a metric set U is an open set, every open set in U is not necessarily a spherical neighborhood. However, it is not difficult to characterize the open sets in U, and we do so in the following theorem, the proof of which is left as an exercise. We note that the empty set is the union of the empty collection of spherical neighborhoods.

Theorem 2.17. A subset G of the metric set U is open iff G is the union of a collection of spherical neighborhoods of U.

Definition 2.18. Let U be a set, and $\mathfrak{S}$ a collection of subsets of U. Then $\mathfrak{S}$ is said to *generate* the collection $\mathfrak{I}$ of subsets of U defined as follows: A subset K of U is an element of $\mathfrak{I}$ iff K is the union of a collection of elements of $\mathfrak{S}$. The collection $\mathfrak{S}$ is said to be a *basis* for the collection which it generates. (Note that, for *any* collection $\mathfrak{S}$, the collection $\mathfrak{I}$ generated by $\mathfrak{S}$ will always contain the empty set as one of its elements.) By $(U, \mathfrak{S})$ we denote the universe U together with the collection $\mathfrak{I}$ of subsets of U generated by $\mathfrak{S}$.

As a result of Theorem 2.17 and Definition 2.18, we see that the open sets of U are *generated* by the collection of all spherical neighborhoods in U. Since the spherical neighborhoods are in turn determined from the metric ρ on U, we thus see that there is an intimate relationship between the metric on U and the open sets of U. It is from this relationship, in fact, that we shall soon arrive at the definition of a metric space. First, however, let us review briefly what we have accomplished thus far.

Given a metric set U, we first use the metric to determine the collection of all spherical neighborhoods in U; these spherical neighborhoods then generate the collection of all open sets in U.

Let us, for convenience, designate by the symbol $\mathfrak{I}$ this collection of all open sets in U. Now, as a result of Theorems 2.9, 2.11, and 2.16 (quoted in parentheses below), the collection $\mathfrak{I}$ has the following properties:

(1) $U \in \mathfrak{I}$ (U is an open set).
(2) $\phi \in \mathfrak{I}$ (ϕ is an open set).
(3) The union of any collection of sets in $\mathfrak{I}$ is a set in $\mathfrak{I}$ (the union of any collection of open sets in U is an open set in U).
(4) The intersection of any finite collection of sets in $\mathfrak{I}$ is a set in $\mathfrak{I}$ (the intersection of any finite collection of open sets in U is an open set in U).

It is worth mentioning, if a slight digression will be allowed here, that if we have *any* set U (whether a metric set or not) together with a collection $\mathfrak{I}$ of open subsets of U which satisfies properties (1) through (4), then we say that $\mathfrak{I}$ is a *topology* for U, and that the set U together with the collection $\mathfrak{I}$ is a *topological space.* Since it is not our intention to investigate abstract topological spaces here, any further discussion of the subject is best postponed to a course in topology. However, we can note that every metric set together with its collection of open sets (as determined by the metric) is a topological space. It is, in fact, a very special kind of topological space, precisely because of the presence of the metric which determines the open sets.

Returning now to our previous discussion of the metric set U, instead of working with the collection $\mathfrak{I}$ of open sets of U, let us back up a step and consider the collection of all spherical neighborhoods in U. If we designate by $\mathfrak{S}$ the collection of all spherical neighborhoods in U, it is easy to verify that $(U, \mathfrak{I})$ is a topological space.

Definition 2.19. A metric set U together with the collection $\mathfrak{I}$ generated by the collection $\mathfrak{S}$ of all spherical neighborhoods in U is a *metric space*, and is denoted by the symbol $(U, \mathfrak{S})$. A subset G of U is open iff G is the union of a collection of sets taken from the collection $\mathfrak{S}$.

The following theorem should now be evident from our previous results.

Theorem 2.20. Let $\mathcal{S}$ be the collection of all spherical neighborhoods in R_1 as determined in Example 2.2. Then $(R_1, \mathcal{S})$ is a metric space.

Merely for easy reference, we now introduce the following notational convention.

Definition 2.21. The phrase "the metric space R_1" always refers to the metric space $(R_1, \mathcal{S})$ of Theorem 2.20; that is, the metric and the spherical neighborhoods are as indicated in Example 2.2.

In dealing with arbitrary metric spaces, we generally use S to denote our universe rather than U. To simplify the notation, we often refer to the metric space S instead of the more proper form $(S, \mathcal{S})$. It should be remembered that whenever we speak of a metric space S, it is understood that we have a given metric on S, and that the collection $\mathcal{S}$ of spherical neighborhoods (and hence the collection $\mathfrak{J}$ of all open sets of S) is determined from this metric.

Now that we have defined a metric space, the question naturally arises as to the distinction between a metric set and a metric space. On the surface, these concepts appear to be about the same, for a metric space is merely a metric set together with the collection $\mathfrak{J}$ generated by its collection $\mathcal{S}$ of spherical neighborhoods (as determined by the metric). However, we have already seen that our primary interest in metric spaces is in the collection $\mathfrak{J}$ of all open subsets of the space. Let us attempt, then, to answer the preceding question by considering instead the following question: Is it possible to define two different metrics on a set (resulting, of course, in distinct metric sets) which ultimately give rise to the same collection $\mathfrak{J}$ of open sets, and hence the same topology for the space? The answer is "yes," and herein lies the distinction; that is, different metric sets may yield what we call equivalent metric spaces, according to the following definition.

Definition 2.22. Two metric spaces are *equivalent* iff they consist of the same set of points (that is, the same universe) and their respective collections of spherical neighborhoods generate the same collection $\mathfrak{J}$ of open sets.

Let S be the universe, and let ρ_1 and ρ_2 be two different metrics on S. Let $\mathcal{S}_1$ be the collection of all spherical neighborhoods in S as determined by the metric ρ_1, and $\mathcal{S}_2$ the collection of all spherical neighborhoods in S as determined by the metric ρ_2. Then, $(S, \mathcal{S}_1)$ and $(S, \mathcal{S}_2)$ are metric spaces.

Theorem 2.23. The metric spaces $(S, \mathcal{S}_1)$ and $(S, \mathcal{S}_2)$ are equivalent iff the following conditions hold true:

(1) Given any spherical neighborhood $U \in \mathcal{S}_1$ and any point $p \in U$, there exists a spherical neighborhood $V \in \mathcal{S}_2$ such that $p \in V \subset U$.

(2) Given any spherical neighborhood $V \in \mathcal{S}_2$ and any point $p \in V$, there exists a spherical neighborhood $U \in \mathcal{S}_1$ such that $p \in U \subset V$.

Proof. Since $(S, \mathcal{S}_1)$ and $(S, \mathcal{S}_2)$ consist of the same universe S, it remains to show, according to Definition 2.22, that the collections $\mathcal{S}_1$ and $\mathcal{S}_2$ generate the same collection $\mathfrak{I}$ of open sets iff conditions (1) and (2) are true. To simplify our notation, let us define $\mathfrak{I}_1$ as the collection of open sets generated by $\mathcal{S}_1$, and $\mathfrak{I}_2$ as the collection of open sets generated by $\mathcal{S}_2$. We thus want to prove that $\mathfrak{I}_1 = \mathfrak{I}_2$ iff conditions (1) and (2) are true. We first prove the "if" (or sufficiency) part of the theorem; that is,

(i) If (1) and (2) are true, then $\mathfrak{I}_1 = \mathfrak{I}_2$.

Since $\mathfrak{I}_1$ and $\mathfrak{I}_2$ are sets, we prove equality by showing inclusion both ways. We shall prove that $\mathfrak{I}_1 \subset \mathfrak{I}_2$, and leave the proof of the reverse inclusion as an exercise. Thus, let W be any member of $\mathfrak{I}_1$. This means that W is an open set in $(S, \mathcal{S}_1)$. If $W = \phi$, then we also have $W \in \mathfrak{I}_2$. If $W \neq \phi$, let p be any point in W. Since W is an open set in $(S, \mathcal{S}_1)$, it follows from Definition 2.5 that there exists $\epsilon > 0$ such that $S(p, \epsilon) \subset W$. But $S(p, \epsilon)$ is a spherical neighborhood in $(S, \mathcal{S}_1)$ and hence belongs to $\mathcal{S}_1$. If we let $U = S(p, \epsilon)$, then we have $p \in U \subset W$, where $U \in \mathcal{S}_1$. Hence by condition (1), there exists V in $\mathcal{S}_2$ such that $p \in V \subset U$, so that $p \in V \subset W$. We have thus shown that if p is any point in W, then there exists a spherical neighborhood $V \in \mathcal{S}_2$ such that $p \in V \subset W$. By Definition 2.5, this says that W is open in $(S, \mathcal{S}_2)$, so that $W \in \mathfrak{I}_2$. Therefore $\mathfrak{I}_1 \subset \mathfrak{I}_2$. This, together with the reverse inclusion, proves (i). We next prove the "only if" (or necessity) part of the theorem; that is,

(ii) If $\mathfrak{J}_1 = \mathfrak{J}_2$, then conditions (1) and (2) are true.

We shall prove that (1) is true, and leave the proof of (2) as an exercise. Let $U \in \mathfrak{S}_1$ and let p be any point of U. We want to show that under the hypothesis $\mathfrak{J}_1 = \mathfrak{J}_2$ there exists $V \in \mathfrak{S}_2$ such that $p \in V \subset U$. Since $U \in \mathfrak{S}_1$, U is a spherical neighborhood in $(S, \mathfrak{S}_1)$, and hence is an open set, by Theorem 2.6; that is, $U \in \mathfrak{J}_1$. Hence, since we are given $\mathfrak{J}_1 = \mathfrak{J}_2$, we must also have $U \in \mathfrak{J}_2$; that is, U is an open set in $(S, \mathfrak{S}_2)$. Then, by Definition 2.5, there exists $\epsilon > 0$ such that $S(p, \epsilon) \subset U$. But $S(p, \epsilon)$ is a spherical neighborhood in $(S, \mathfrak{S}_2)$, and hence belongs to $\mathfrak{S}_2$. Letting $V = S(p, \epsilon)$, we thus have $p \in V \subset U$, which is the desired result. QED

An interesting application of Theorem 2.23 may be illustrated in the plane R_2. Let us denote by ρ_1 the metric on R_2 defined at the end of Section 1; that is, if $p_1 = (x_1, y_1)$ and $p_2 = (x_2, y_2)$ are any two points in R_2, then

$$\rho_1(p_1, p_2) = [(x_1 - x_2)^2 + (y_1 - y_2)^2]^{1/2}$$

Let $\mathfrak{S}_1$ be the collection of all spherical neighborhoods in R_2 as determined by the metric ρ_1, so that $(R_2, \mathfrak{S}_1)$ is a metric space. We have already seen that for any point p in R_2 and any $\epsilon > 0$, the spherical neighborhood $S(p, \epsilon)$, which is a typical element of $\mathfrak{S}_1$, is the interior of a circle of radius ϵ and centered at p.

Now let us appeal for a moment to the student's geometric intuition. If we draw any circle in the plane and choose any point p in the interior of the circle (p may be very "close" to, but not on, the boundary of the circle), it seems intuitively clear that we can draw a square with sides parallel to the coordinate axes such that the square has p as its center and is entirely contained in the interior of the circle. In fact, we could do this by first drawing a smaller circle centered at p and entirely contained in the given circle, then inscribing the desired square in this smaller circle. Next, suppose we reverse this process. If we draw a square in the plane with sides parallel to the coordinate axes, and choose any point p in the interior of the square, it seems equally clear that we can draw a circle centered at p and entirely contained in the interior of the square. Again, we could accomplish this by first drawing a smaller square centered at p and entirely contained in the given square, then inscribing a circle in this smaller square.

Let us suppose now that we are able to define another metric ρ_2 on R_2 in such a way that a typical spherical neighborhood of a point p in R_2 (and hence a typical element in the collection $\mathcal{S}_2$ of all spherical neighborhoods) is the interior of a square centered at p and with sides parallel to the coordinate axes. By virtue of our preceding intuitive argument and Theorem 2.23, it appears that $(R_2, \mathcal{S}_1)$ and $(R_2, \mathcal{S}_2)$ must be equivalent metric spaces. If so, this implies that the collection $\mathcal{S}_1$ of "open circles" and the collection $\mathcal{S}_2$ of "open squares" generate the same collection $\mathfrak{J}$ of open sets in R_2.

Returning from the hypothetical to the concrete, we now make the rather startling assertion embodied in the following example.

Example 2.24. There does exist a metric ρ_2 on R_2 such that a typical spherical neighborhood of a point p in R_2 is the interior of a square centered at p and with sides parallel to the coordinate axes. To show this, we must exhibit the metric ρ_2 and then verify our assertion. Let $p_1 = (x_1, y_1)$ and $p_2 = (x_2, y_2)$ be any two points in R_2, and define ρ_2 as follows:

$$\rho_2(p_1, p_2) = \max [|x_1 - x_2|, |y_1 - y_2|]$$

We leave as an exercise the verification that ρ_2 is a metric on R_2. It remains then to determine the nature of a typical spherical neighborhood. Since a particular spherical neighborhood of a point $p = (h, k)$ in R_2 is merely a translate of a corresponding spherical neighborhood of the point $(0, 0)$, it suffices to consider a spherical neighborhood of the origin. In particular, let us first determine the "unit sphere" in $(R_2, \mathcal{S}_2)$; that is, the set of all points in R_2 which are a distance of one unit (according to the metric ρ_2) from the origin. We thus want to determine the set

$$\{p : p \in R_2, \rho_2[p, (0, 0)] = 1\}$$

From the definition of ρ_2 this is the same as the set

$$\{(x, y) : \max [|x - 0|, |y - 0|] = 1\}$$

or

$$\{(x, y) : \max [|x|, |y|] = 1\}$$

Note that all points (x, y) such that $y = \pm 1$ and $-1 \leq x \leq 1$ satisfy $|y| = 1$, and hence are on the unit sphere. Similarly, all points (x, y) such that $x = \pm 1$ and $-1 \leq y \leq 1$ satisfy $|x| = 1$, and are also on the unit sphere. Thus the unit sphere is the square with vertices at the points $(1, 1)$, $(1, -1)$, $(-1, -1)$, $(-1, 1)$. The interior of this square is a typical neighborhood of (0, 0). Its center is at the origin, and its sides are of length 2 and parallel to the coordinate axes. Generalizing this result, for any point p in R_2 and any $\epsilon > 0$, the spherical neighborhood $S(p, \epsilon)$ in $\mathcal{S}_2$ is the interior of a square centered at p with sides of length 2ϵ parallel to the coordinate axes. This concludes Example 2.24.

For the sake of completeness, we now state the equivalence of these metric spaces as a theorem. The method of proof is suggested by our earlier argument based on geometric intuition. For this reason, we merely sketch the proof and leave the details as an exercise.

Theorem 2.25. The metric spaces $(R_2, \mathcal{S}_1)$ and $(R_2, \mathcal{S}_2)$ are equivalent.

Proof. We must prove that conditions (1) and (2) of Theorem 2.23 hold true. To prove (1), let U be any element in $\mathcal{S}_1$, and p any point in U. Since U is an open set in $(R_2, \mathcal{S}_1)$, there exists $\epsilon_1 > 0$ such that $S(p, \epsilon_1) \subset U$. Let $W = S(p, \epsilon_1)$. Then $W \in \mathcal{S}_1$; that is, W is the interior of a circle centered at p and with radius ϵ_1, so that $p \in W \subset U$. Now we proceed to inscribe a square in the circle W. Let $\epsilon_2 = \epsilon_1/\sqrt{2}$, and define $V = \{q : q \in R_2, \rho_2(p, q) < \epsilon_2\}$. It can be shown that $V \in \mathcal{S}_2$ and $V \subset W$. Therefore $p \in V \subset U$. To prove condition (2) of Theorem 2.23, let V be any element in $\mathcal{S}_2$, and p any point in V. Since V is an open set in $(R_2, \mathcal{S}_2)$, there exists $\epsilon > 0$ such that $S(p, \epsilon) \subset V$. Let $W = S(p, \epsilon)$. Then $W \in \mathcal{S}_2$; that is, W is the interior of a square centered at p with sides of length 2ϵ parallel to the coordinate axes, so that $p \in W \subset V$. Now we inscribe a circle in the square W. Define $U = \{q : q \in R_2, \rho_1(p, q) < \epsilon\}$. It can be shown that $U \in \mathcal{S}_1$ and $U \subset W$. Therefore $p \in U \subset V$.
QED

At the beginning of this section, we discussed the possibility of extending many of the important properties of the metric set R_1 to arbitrary metric sets. We began with the concept of open sets, which ultimately led to the definition of a metric space. By virtue of Theorem 2.20, our earlier results in R_1 which depend solely on the metric or on the open sets may be considered as properties of the metric space R_1. Our goal now is to generalize these properties to arbitrary metric spaces. We conclude this section with a discussion of boundedness, and consider additional properties in subsequent sections.

Definition 2.26. A subset A of a metric space S is *bounded* with respect to the metric ρ of S iff A is empty or there exists a real number $\alpha > 0$ such that $\rho(x, y) \leq \alpha$ for every $x, y \in A$.

Note that if we choose S to be the metric space R_1, this definition is equivalent to our original definition of a bounded set in R_1 (see Problem 29 of Chapter I). We thus have extended the concept of boundedness from the metric space R_1 to general metric spaces.

It is an obvious and seemingly uninteresting fact that the metric space R_1 is not bounded. However, one of the fruitful results of our general approach to metric spaces is that we may have a metric space S, with metric ρ, such that S (and hence every subset of S) is bounded with respect to ρ. Such a metric is called a bounded metric for S.

Definition 2.27. A metric ρ for the metric space S is said to be a *bounded metric* for S iff S is bounded with respect to this metric ρ.

We close this section with a very important theorem concerning bounded metrics.

Theorem 2.28. Every metric space is equivalent to a metric space with a bounded metric.

Proof. Let S be a metric space with metric ρ. Define the mapping $\rho_1 \colon S \times S \to R_1$ as follows: For every $(x, y) \in S \times S$,

$$\rho_1(x, y) = \frac{\rho(x, y)}{1 + \rho(x, y)}$$

We must show first that ρ_1 is a metric on S. The first three conditions for a metric follow easily from the fact that ρ is a metric on S. To prove the triangle inequality, let x, y, and z be any three points of S, and let us suppose that $x \neq z$. We sketch the fundamental chain of inequalities, and suggest that the student verify each inequality in the chain:

$$\begin{aligned}\rho_1(x, y) + \rho_1(y, z) &= \frac{\rho(x, y)}{1 + \rho(x, y)} + \frac{\rho(y, z)}{1 + \rho(y, z)} \\ &\geq \frac{\rho(x, y)}{1 + \rho(x, y) + \rho(y, z)} + \frac{\rho(y, z)}{1 + \rho(x, y) + \rho(y, z)} \\ &= \frac{\rho(x, y) + \rho(y, z)}{1 + \rho(x, y) + \rho(y, z)}\end{aligned}$$

$$= \frac{1}{1 + \dfrac{1}{\rho(x, y) + \rho(y, z)}}$$

$$\geq \frac{1}{1 + \dfrac{1}{\rho(x, z)}}$$

$$= \rho_1(x, z)$$

We thus have shown that

$$\rho_1(x, z) \leq \rho_1(x, y) + \rho_1(y, z)$$

for any three points x, y, and z with $x \neq z$. But when $x = z$, the inequality follows immediately. Therefore ρ_1 is a metric on S. Let us denote by $\mathfrak{S}_1$ the collection of all spherical neighborhoods in S determined by the metric ρ_1, and by S_1 the metric space $(S, \mathfrak{S}_1)$. Then ρ_1 is a bounded metric for S_1, since for every pair of points $x, y \in S$, it follows from the definition of ρ_1 that $\rho_1(x, y) < 1$.

It remains to prove that the metric spaces S and S_1 are equivalent. We do this by showing that each space has the same collection of spherical neighborhoods. Thus let U be a typical spherical neighborhood in the metric space S; that is, for any point p in the set S and $\epsilon > 0$, let $U = S(p, \epsilon)$. Our aim is to find a spherical neighborhood V of p in the metric space S_1 such that $U = V$. Since $U = S(p, \epsilon)$ is determined by the metric ρ, it follows that for every $x \in U$,

(1) $$\rho(p, x) < \epsilon$$

(2) $$\rho(p, x) + \epsilon \cdot \rho(p, x) < \epsilon + \epsilon \cdot \rho(p, x)$$

(3) $$\rho(p, x)(1 + \epsilon) < \epsilon[1 + \rho(p, x)]$$

(4) $$\frac{\rho(p, x)}{1 + \rho(p, x)} < \frac{\epsilon}{1 + \epsilon}$$

From the definition of ρ_1, we see that (4) may be rewritten as

(5) $$\rho_1(p, x) < \frac{\epsilon}{1 + \epsilon}$$

Now let us make the following definitions:

(6) $$\epsilon_1 = \frac{\epsilon}{1 + \epsilon}$$

(7) $$V = \{x : x \in S, \rho_1(p, x) < \epsilon_1\}$$

Then V is a spherical neighborhood of p in the metric space S_1; furthermore, since (5) is true for every $x \in U$, it follows that $U \subset V$. On the other hand, for every $x \in V$, it is evident from the definition of ϵ_1 that (5) is true. We can then reverse steps (5) through (1) to get $\rho(p, x) < \epsilon$, so that $x \in U$. Hence $V \subset U$, which, combined with the inclusion $U \subset V$ determined previously, yields $U = V$. We have now shown that every spherical neighborhood in the metric space S is also a spherical neighborhood in S_1.

To complete the proof, we must show that every spherical neighborhood in S_1 is also a spherical neighborhood in S. By a judicious choice of notation here, we may simplify our work by using some of the results already established. Thus let V be a typical spherical neighborhood in the metric space S_1; that is, for any point p in the set S and any ϵ_1 such that $0 < \epsilon_1 < 1$, let $V = S(p, \epsilon_1)$. Recall that S_1 is a bounded metric space, and, in fact, $\rho_1(p, x) < 1$ for every x in S_1, this explains our restriction on ϵ_1. Now define ϵ by

(8) $$\epsilon = \frac{\epsilon_1}{1 - \epsilon_1}$$

Since $0 < \epsilon_1 < 1$, we see that $\epsilon > 0$; furthermore, it is easy to verify that the ϵ and ϵ_1 of (8) also satisfy (6). Thus, if we define U by

(9) $$U = \{x : x \in S, \rho(p, x) < \epsilon\}$$

and note that U is a spherical neighborhood of p in the metric space S, we can employ the same type of argument as used in the first half of the proof to show that $V = U$. We leave the details as an exercise. QED

PROBLEMS

5. Prove Theorem 2.12.
6. Prove Theorem 2.15.
7. Prove Theorem 2.16.
8. Prove Theorem 2.17.
9. Complete the proof of Theorem 2.23 as follows:
In (i), prove that $\mathfrak{J}_2 \subset \mathfrak{J}_1$.
In (ii), prove that (2) is true.
10. Prove that the mapping ρ_2 defined in Example 2.24 is a metric on R_2.

11. Fill in the details in the proof of Theorem 2.25 at the two points where the assertion "It can be shown that . . ." occurs.
12. Furnish the details to prove that $V = U$ as asserted in the second half of the proof of Theorem 2.28.
13. Prove that every subset A of a metric space S is a metric space. Furthermore, show that if S is bounded, then A is bounded. Conclude that in this case, the set A when considered as the universe is a metric space that is both closed and bounded.
14. Let S be a metric space, and $\mathfrak{I}$ the collection of all open subsets of S. Let $\mathfrak{K}$ be the collection of sets generated by $\mathfrak{I}$. Prove $\mathfrak{K} = \mathfrak{I}$.

3. Sequentially compact sets in metric spaces

In this section, we continue to extend some of our earlier results for the metric space R_1 to general metric spaces. The proofs of many of the theorems may be obtained from the proofs of the corresponding theorems for R_1 merely by substituting the general metric $\rho(x, y)$ for the R_1 metric $|x - y|$. Consequently the proofs of such theorems are usually omitted, with the understanding that the student will supply the details.

Bear in mind that whenever we speak of a metric space S, we imply the existence of a metric ρ on S. We begin with a brief discussion of sequences.

Definition 3.1. Let S be a metric space with metric ρ. A sequence $\langle p_n \rangle$ of points of S *converges* to the point p of S iff given any $\epsilon > 0$, there exists a positive integer N such that if $n > N$, then $\rho(p_n, p) < \epsilon$. A sequence $\langle p_n \rangle$ of points of S is said to be *convergent* iff there is a point p of S to which $\langle p_n \rangle$ converges. We say that p is the *limit* of the sequence $\langle p_n \rangle$, and write $\lim p_n = p$.

It is fairly easy to show (as we did in R_1) that in a metric space a convergent sequence of points has a unique limit. (See Problem 16 in this chapter.) Moreover, this property of uniqueness of limits is related to another important property of metric spaces, which is known as the Hausdorff property.

Theorem 3.2. Every metric space S has the Hausdorff property; that is, given any two distinct points p, q of S, there exist disjoint open sets U and V of S such that $p \in U$, $q \in V$.

Proof. Since $p \neq q$, it follows that $\rho(p, q) > 0$. Let r be any real number such that $0 < 2r < \rho(p, q)$. Then the spherical neighborhoods $S(p, r)$ and $S(q, r)$ are disjoint open sets of S containing p and q, respectively. QED

The Hausdorff property may be classified as a "separation" property. Roughly speaking, it assures us that the collection of open sets in the metric space S is sufficiently large so that we can separate distinct points of S by means of open sets. Note that we used the Hausdorff property of R_1 in our proof of Theorem 1.7 of Chapter IV.

Our previous definition for Cauchy sequences generalizes at once to arbitrary metric spaces.

Definition 3.3. A sequence $\langle p_n \rangle$ of points in a metric space is said to be a *Cauchy sequence* iff given any $\epsilon > 0$, there exists a positive integer N such that if h and k are any two integers with $h > N$, $k > N$, then $\rho(p_h, p_k) < \epsilon$.

Corresponding to Theorem 1.10 of Chapter IV, we now have the following immediate result.

Theorem 3.4. Every convergent sequence in a metric space is a Cauchy sequence.

Following the pattern of our results in Chapter IV, we might hope at this point to prove the converse of Theorem 3.4, and hence derive a "Cauchy criterion" for convergence in general metric spaces. Unfortunately, the converse of Theorem 3.4 is not true in general metric spaces as shown by the following example.

Example 3.5. Let S be the set of real numbers defined by $S = \{x: 0 < x \leq 1\}$ and let ρ be the usual R_1 metric; that is, for any two elements $x, y \in S$, $\rho(x, y) = |x - y|$. Then S together with the collection of sets generated by the collection of spherical neighborhoods of S (as determined by ρ) is a metric space. Consider the sequence $\langle p_n \rangle$ of points of S defined by $p_n = 1/n$ for each positive integer n. This sequence is a Cauchy sequence in S, but it is not a convergent sequence in S. Note that since $0 \notin S$, there is no point of S to which the sequence converges.

We now define and discuss briefly the concept of a cluster point of a subset of a metric space.

Definition 3.6. Let E be a subset of a metric space S. Then a point p of S is said to be a *cluster point* (or an *accumulation point*)

of E iff there exists a sequence $\langle p_n \rangle$ of distinct points of E such that $\langle p_n \rangle$ converges to p.

The next theorem is a generalization of Theorem 4.5 of Chapter IV.

Theorem 3.7. Let E be a subset of a metric space S. Then a point p of S is a cluster point of E iff given any $\epsilon > 0$, there exists a point q of E such that $q \neq p$ and $q \in S(p, \epsilon)$.

As an immediate consequence of Theorem 3.7, we have the following.

Corollary 3.8. Let E be a subset of a metric space S. Then a point p of S is not a cluster point of E iff there is an $\epsilon > 0$ such that the spherical neighborhood $S(p, \epsilon)$ contains no point of E distinct from p. (The point p itself may or may not belong to E.)

As in R_1, we may now characterize closed sets in a metric space in terms of cluster points. We do so by first defining the closure of a set.

Definition 3.9. Let E be a subset of a metric space S, and denote by E' the set of all cluster points of E. Then the *closure* of the set E (written $\bar{E}$) is the subset of S defined by $\bar{E} = E \cup E'$.

The properties of closure as proved for R_1 in Chapter IV (note in particular Theorem 4.17) are equally true for general metric spaces; thus we shall use them without further reference.

Theorem 3.10. A set E in a metric space is closed iff $E = \bar{E}$.

We are finally ready to discuss sequential compactness in metric spaces. We shall use precisely Definition 4.19 of Chapter IV.

Definition 3.11. A subset H of a metric space S is *sequentially compact* iff given any sequence $\langle p_n \rangle$ of points of H, there exists a subsequence $\langle p_{n_i} \rangle$ of $\langle p_n \rangle$ which converges to a point of H.

Theorem 3.12. A subset H of a metric space S is sequentially compact iff every infinite subset K of H has at least one cluster point in H.

Proof. We first show that if the condition holds, then H is sequentially compact. Let $\langle p_n \rangle$ be any sequence of points of H. If $\langle p_n \rangle$ contains only a finite number of distinct terms, then there is a point p of H and a subsequence $\langle p_{n_i} \rangle$ of $\langle p_n \rangle$ such that $p_{n_i} = p$ for every $i \in I_0$. Clearly, the subsequence $\langle p_{n_i} \rangle$ of $\langle p_n \rangle$ converges to p. We consider next the case where the sequence $\langle p_n \rangle$ has infinitely many distinct terms. Define $K = \bigcup_{n \in I_0} \{p_n\}$. Then K is an infinite subset of H. By the given condition, K has a cluster point p in H, and this means that there exists a subsequence $\langle p_{n_i}^* \rangle$ of $\langle p_n \rangle$ which converges to the point p of H. This completes the proof that if the condition holds, then H is sequentially compact.

We suppose next that H is sequentially compact, and prove that the given condition holds. Let K be any infinite subset of H; then K contains a sequence $\langle p_n \rangle$ of distinct points of H. Since H is sequentially compact, there is a point p of H and a subsequence $\langle p_{n_i} \rangle$ of $\langle p_n \rangle$ which converges to p. Since we have found a sequence $\langle p_{n_i} \rangle$ of distinct points of K which converges to the point p of H, the point p of H is a cluster point of K. Therefore the given condition holds. QED

Corollary 3.13. A subset H of a metric space S is not sequentially compact iff there exists an infinite subset K of H which has no cluster point in H.

Corollary 3.14. A subset H of a metric space S is not sequentially compact iff there exists a sequence $\langle p_n \rangle$ of distinct points of H such that no subsequence of $\langle p_n \rangle$ converges to a point of H.

Corollary 3.15. The set $S = \{x: 0 < x \leq 1\}$ described in Example 3.5 is not sequentially compact.

We have seen that in the metric space R_1, a set is sequentially compact iff it is closed and bounded. It would be natural to hope that this result might hold for general metric spaces, but unfortunately it does not. We shall show, in fact, that closedness and boundedness of a set in a metric space are necessary, but not sufficient, conditions for sequential compactness.

Example 3.16. Let T be the metric space consisting of all points of R_1 except zero, and with the usual metric for R_1, and define S as in Corollary 3.15. The set S is clearly a closed and bounded subset of T. However, S is not sequentially compact, by Corollary 3.15.

The result of this example is sufficiently important to be stated as a theorem.

Theorem 3.17. It is not true that every closed and bounded subset of a metric space is sequentially compact.

The following lemmas are useful in proving that every sequentially compact subset of a metric space is closed and bounded. They are also fundamental to our discussion of separability. The statements of these lemmas require the idea of an ϵ-net, which we now define.

Definition 3.18. Let A be a nonempty set, and $\epsilon > 0$ a real number. By an *ϵ-net* in the set A we mean a finite collection of points $p_1, p_2, \ldots, p_N$ of A such that given any point $q \in A$, there exists an integer i, $1 \leq i \leq N$, such that $\rho(q, p_i) < \epsilon$.

Lemma 3.19. Let A be a nonempty finite set, and let $\epsilon > 0$ be given. Then A has an ϵ-net.

The proof is left as an exercise.

Definition 3.20. Let $\epsilon > 0$ be given. Then a set H is said to be *ϵ-isolated* iff given any two distinct points x, y of H, $\rho(x, y) \geq \epsilon$.

Lemma 3.21. Let $\epsilon > 0$ be given, and let A be a nonempty set which has no ϵ-net. Let n be a positive integer. Then there exists a subset K of A with the following properties:

(1) K consists of exactly n points.

(2) K is ϵ-isolated.

Proof. It is immediate that the lemma is true when $n = 1$, since the set $K = \{p\}$, where p is any point of the nonempty set A, satisfies both (1) and (2).

Let H be the set of all positive integers for which the lemma is false. If H is empty, then the proof of the lemma is complete. We may thus assume that H is nonempty. By the Well-Ordering Axiom, H must contain a smallest element j, and by the first paragraph of the proof, $j > 1$. The lemma must then be true for the positive integer $j - 1$. This tells us that we can find a subset K of A satisfying the following conditions:

(a) K consists of exactly $j - 1$ points.

(b) K is ϵ-isolated.

There must exist a point p of A such that $\rho(p, q) \geq \epsilon$ for every q in K. This follows at once from the fact that A has no ϵ-net; in particular, K is not an ϵ-net for A. Consider the set $M = K \cup \{p\}$, and note that this set satisfies (1) and (2) for $n = j$. This contradiction completes the proof of the lemma. QED

Lemma 3.22. Let A be a nonempty sequentially compact set, and let $\epsilon > 0$ be given. Then A has an ϵ-net.

Proof. Suppose the lemma is false. Then there exists an $\epsilon > 0$ such that A does not have an ϵ-net. For each positive integer n, let $\mathfrak{M}_n$ be the collection of all subsets K of A satisfying the following conditions:

(1) K consists of exactly n points.

(2) K is ϵ-isolated.

By Lemma 3.21, each of the sets $\mathfrak{M}_n$ is nonempty. By the Axiom of Choice for Sequences, there must exist, for each positive integer n, an element K_n of $\mathfrak{M}_n$. Note that the set K_n satisfies conditions (1) and (2).

We now define a sequence of subsets $\langle H_i \rangle$ of A with the following properties:

(*a*) H_i consists of exactly i points.

(*b*) $H_i \subset H_{i+1}$ for each $i \in I_*$.

(*c*) Each H_i is $\epsilon/2$-isolated.

The sequence $\langle H_i \rangle$ is obtained from the sequence $\langle K_n \rangle$. We observe that for each n, the set K_n consists of exactly n points, so that we may write the sets K_n as follows:

$$
\begin{aligned}
K_1 &= \{p_{1,1}\} \\
K_2 &= \{p_{2,1}\} \cup \{p_{2,2}\} \\
K_3 &= \{p_{3,1}\} \cup \{p_{3,2}\} \cup \{p_{3,3}\} \\
&\;\;\vdots \qquad\qquad \vdots \\
K_n &= \{p_{n,1}\} \cup \{p_{n,2}\} \cup \{p_{n,3}\} \cup \cdots \cup \{p_{n,n}\} \\
&\;\;\vdots \qquad\qquad \vdots
\end{aligned}
$$

Define $q_1 = p_{1,1}$, and let $H_1 = \{q_1\}$. We note that $\rho(p_{2,1}, p_{2,2}) \geq \epsilon$. A simple application of the triangle inequality shows that we must have either $\rho(q_1, p_{2,1}) \geq \epsilon/2$ or $\rho(q_1, p_{2,2}) \geq \epsilon/2$. Define q_2 to be the element of K_2 with smallest second subscript such that this condition is satisfied. Define $H_2 = H_1 \cup \{q_2\}$. Note that H_2 satisfies conditions (*a*), (*b*), and (*c*) for $i = 2$. We complete the definition of the sequence $\langle H_i \rangle$ by an inductive argument. Suppose that the sets $H_1, H_2, \ldots, H_k$ have been defined to satisfy the following conditions:

(*a*) H_i consists of exactly i points, for $i = 1, 2, \ldots, k$.
(*b*) $H_1 \subset H_2 \subset \ldots \subset H_k$.
(*c*) H_i is $\epsilon/2$-isolated, for $i = 1, 2, 3, \ldots, k$.
(*d*) $H_k = \{q_1\} \cup \{q_2\} \cup \ldots \cup \{q_k\}$.

We now recall that the set K_{k+1} may be written as

$$K_{k+1} = \{p_{k+1,1}\} \cup \{p_{k+1,2}\} \cup \ldots \cup \{p_{k+1,k+1}\}$$

Let x and y be any two distinct points of K_{k+1}. Since K_{k+1} is ϵ-isolated, we must have $\rho(x, y) \geq \epsilon$. A simple application of the triangle inequality shows that both the conditions

$$\rho(x, q_i) < \frac{\epsilon}{2} \qquad \text{and} \qquad \rho(y, q_i) < \frac{\epsilon}{2}$$

cannot be true, for every $i = 1, 2, \ldots, k$. This means that no two distinct points of K_{k+1} can each be within a distance $\epsilon/2$ of the same point of H_k. However, there are $k + 1$ points in K_{k+1}, and only k points in H_k. Therefore, there must exist at least one point z of K_{k+1} such that $\rho(z, q_i) \geq \epsilon/2$ for every point q_i of H_k. Define q_{k+1} to be the point of K_{k+1} with smallest second subscript which satisfies this last condition. The set $H_{k+1} = H_k \cup \{q_{k+1}\}$ must then satisfy the preceding conditions (*a*), (*b*), (*c*), and (*d*). This completes the inductive definition of the sequence $\langle H_i \rangle$.

Define $H = \bigcup_{i \in I_0} H_i$. Then H is an infinite set. By Theorem 1.22 of Chapter III, H contains a sequence $\langle p_n \rangle$ of distinct points. This sequence has the property that $\rho(p_i, p_j) \geq \epsilon/2$ for every pair of distinct positive integers i and j. It follows at once that the sequence $\langle p_n \rangle$ is uniformly isolated. Consequently, this sequence has no convergent subsequence. This contradicts the hypothesis that A is sequentially compact. QED

Theorem 3.23. Every sequentially compact subset A of a metric space S is both closed and bounded.

Proof. We leave as an exercise the proof that A is closed. To prove that A is bounded, pick any $\epsilon > 0$. By Lemma 3.22, there exists an ϵ-net $p_1, p_2, \ldots, p_N$ of points of A. Let p and q be arbitrary points of A. Then there exist integers i and j, $1 \leq i \leq N$, $1 \leq j \leq N$, such that $\rho(p, p_i) < \epsilon$ and $\rho(q, p_j) < \epsilon$. Define $B = \sup \{\rho(p_i, p_j): 1 \leq i \leq N, 1 \leq j \leq N\}$. We note that

$$\begin{aligned} \rho(p, q) &\leq \rho(p, p_i) + \rho(p_i, p_j) + \rho(p_j, q) \\ &< 2\epsilon + \rho(p_i, p_j) \\ &\leq 2\epsilon + B \end{aligned}$$

It follows that for *any* two points p and q of A, the distance between p and q is less than the real number $2\epsilon + B$. Therefore A is bounded. QED

PROBLEMS

15. Prove that if p and q are points of a metric space such that given any $\epsilon > 0$, $\rho(p, q) < \epsilon$, then $p = q$.
16. Prove, in the following two ways, that a convergent sequence $\langle p_n \rangle$ of points in a metric space S has a unique limit p.
(*a*) Follow the general pattern of the proof of Theorem 1.7, Chapter IV, using the Hausdorff property of S and Theorem 3.3.
(*b*) Suppose that $p, q \in S$, with $\lim p_n = p$ and $\lim p_n = q$. Use Definition 3.1, the triangle property of the metric ρ, and Problem 15 above to prove that $p = q$.
17. Prove that if p and q are points of a metric space such that $0 < 2r < \rho(p, q)$, then the spherical neighborhoods $S(p, r)$ and $S(q, r)$ are disjoint open sets.
18. Prove Theorem 3.4.
19. Prove: A convergent sequence $\langle p_n \rangle$ of points of a metric space is bounded; that is, there exists a real number $K > 0$ such that $\rho(p_i, p_j) \leq K$ for every i, j.
20. Prove that every subsequence of a convergent sequence is convergent, and converges to the same point as the given sequence.
21. Prove Theorem 3.7.
22. Prove Corollary 3.8.
23. Prove that for every subset E of a metric space S the set E' is a closed subset of S.
24. Prove Corollary 3.13.
25. Prove Corollary 3.14.
26. Prove Corollary 3.15.
27. Prove that every sequence $\langle p_n \rangle$ of distinct points of a sequentially compact set contains a subsequence which is a Cauchy sequence.
28. Prove that every sequentially compact set E is *complete*; that is, a

sequence of points $\langle p_n \rangle$ of E converges to a point of E iff $\langle p_n \rangle$ is a Cauchy sequence.

29. Prove that every sequentially compact subset of a metric space is closed.

30. Prove Lemma 3.19, and state its contrapositive.

31. Carry out the "simple applications of the triangle inequality" in the proof of Lemma 3.22.

32. Prove that every closed subset of a sequentially compact metric space is sequentially compact.

33. Let H be the set of all rational numbers in the metric space R_1. Prove that $\overline{H} = R_1$. (*Hint:* See Theorem 3.14 of Chapter III.)

34. Let H be a subset of a metric space S, and p a point of S. Then $p \in \overline{H}$ iff given any $\epsilon > 0$, there exists a point q of H such that $\rho(p, q) < \epsilon$.

4. *Separability and compactness; Lindelöf and Cantor Product Theorems*

Consider the metric space R_1 and the countable subset H of R_1 consisting of all rational numbers. It follows from Problem 33 that $\overline{H} = R_1$. Thus there exists a countable subset H of the metric space R_1 such that every point of R_1 is either a point of H or a cluster point of H. For this reason, R_1 is said to be separable.

Definition 4.1. A metric space S is said to be *separable* iff there exists a countable subset H of S such that $\overline{H} = S$.

Theorem 4.2. Every sequentially compact metric space S is separable.

Proof. If S is empty, then define $H = \phi$. It follows that H is countable, and $\overline{H} = S$. Hence S is separable. Assume that S is not empty. For each positive integer n, there exists, by Lemma 3.22, a finite subset H_n of S which is a $1/n$-net in S. Define $H = \bigcup_{n \in I_0} H_n$, and note that H is a countable set. The proof of the theorem will be complete if we show that $\overline{H} = S$. Since $\overline{H} \subset S$, it is sufficient to show that $S \subset \overline{H}$. To this end, let p be any point of S, and let $\epsilon > 0$ be given. There exists an integer N such that $1/N < \epsilon$. Now the subset H_N of H is a $1/N$-net in S. Accordingly, there must exist a point q in H_N such that $\rho(p, q) < 1/N < \epsilon$. Therefore by Problem 34, $p \in \overline{H}$. QED

Our definition of a compact subset of a metric space S is an immediate generalization of Definition 4.14, Chapter III. We also use the notion of an open covering of a subset of a metric space. For this idea, the student is referred to Definition 4.13 of Chapter III.

Definition 4.3. A subset K of a metric space S is said to be *compact* iff every open covering of K contains a finite open subcovering of K.

We now specialize this definition to the case where the given open covering of K is countable; that is, the open covering $\{G_\alpha\}$ consists of a countable collection of open sets. Then, we write the open covering as $\{G_n\}$.

Definition 4.4. A subset K of a metric space S is said to be *countably compact* iff every countable open covering of K contains a finite open subcovering of K.

The following theorem is immediate.

Theorem 4.5. Every compact subset K of a metric space S is a countably compact subset of S.

Perhaps the most important property of separable metric spaces is the fact that every open covering of such a space has a countable open subcovering. The proof of this important result is based on the fact that every separable metric space has a countable basis. (See Definition 2.18.)

Theorem 4.6. Every separable metric space S has a countable basis.

Proof. Let $(S, \mathcal{S})$ be the given metric space. Since S is separable, there exists a countable subset H of S such that $\overline{H} = S$. Since H is countable, there exists a mapping $f\colon I_o \xrightarrow{\text{onto}} H$. For each $n \in I_o$, define $p_n = f(n)$. Then $H = \bigcup_{n \in I_o} \{p_n\}$. Define the set T to consist of all pairs (h, k), where h and k are positive integers. We know that the set T is countable, by Theorems 1.10 and 1.24 of Chapter III. Now, consider the collection of spherical neighborhoods $\{S(p_i, 1/j)\}$, and define a mapping g of the set T onto this collection by $g[(i, j)] = S(p_i, 1/j)$ for each element (i, j) of T. It follows from Theorem 1.7 of Chapter III that the collection $\{S(p_i, 1/j)\}$ is countable. Denote this collection by $\mathfrak{J}$. The proof of our theorem will be complete if we show that $(S, \mathcal{S})$ and $(S, \mathfrak{I})$ are equivalent metric spaces, and for this we make use of Theorem 2.23.

Let U be any element of $\mathcal{S}$. We recall that $\mathcal{S}$ is a collection of open subsets of S which generates the collection of all open subsets of S. Thus U is an open subset of S. Let p be any point of U. There exists an $\epsilon > 0$ such that $S(p, \epsilon) \subset U$. There exists an integer n such that p_n is an element of $S(p, \epsilon/2)$. There exists an integer j such that $1/j < \epsilon/2$; hence, $S(p_n, 1/j) \subset S(p_n, \epsilon/2)$. The spherical neighborhood $S(p_n, 1/j)$ is an element of $\mathfrak{I}$. We leave it as an exercise for the student to show that

$$p \in S\left(p_n, \frac{1}{j}\right) \subset U$$

This proves that condition (1) of Theorem 2.23 holds. We leave as an exercise the proof that condition (2) of Theorem 2.23 holds.

QED

Theorem 4.7 (Lindelöf Theorem). Let S be any metric space which has a countable basis $\{U_n\}$, and let A be any subset of S. Then every open covering of A has a countable open subcovering.

Proof. Let $\{G_\alpha\}$ be any open covering of A. Pick a particular G_α and denote it by G_0. For each positive integer n, define $\mathfrak{M}_n$ in the following way: If there exists any G_α which contains U_n, define $\mathfrak{M}_n$ to be the collection of all such G_α; if there exists no G_α which contains U_n, define $\mathfrak{M}_n$ to consist of the single set G_0. Note that every $\mathfrak{M}_n$ is nonempty. By the Axiom of Choice for Sequences, there exists a sequence $\langle G_n \rangle$ such that $G_n \in \mathfrak{M}_n$ for each $n \in I_o$.

Let $p \in A$. Since $\{G_\alpha\}$ is an open covering of A, there exists an index α_0 such that $p \in G_{\alpha_0}$. From the fact that $\{U_n\}$ is a basis for the metric space S, and the fact that G_{α_0} is open, we see that there must exist an integer n_0 such that

$$p \in U_{n_0} \subset G_\alpha$$

This means that there is at least one G_α containing U_{n_0}. Accordingly,

$$p \in U_{n_0} \subset G_{n_0}$$

It follows at once that $A \subset \bigcup_{n \in I_o} G_n$. QED

Corollary 4.8. Every open covering of a separable metric space has a countable open subcovering.

Corollary 4.9. Every open covering of a sequentially compact metric space has a countable open subcovering.

Theorem 4.10. Every countably compact metric space S is sequentially compact.

Proof. Suppose that S is not sequentially compact. Then, by Corollary 3.14, there exists a sequence $\langle p_n \rangle$ of distinct points of S such that no subsequence of $\langle p_n \rangle$ converges to a point of S. Define $H = \bigcup_{n \in I_o} \{p_n\}$, and note that for each $n \in I_o$, p_n is not a cluster point of H. Define $\mathfrak{M}_n$, for each $n \in I_o$, as the collection of all neighborhoods V_α of p_n such that $V_\alpha \cap H = \{p_n\}$. Note that for each $n \in I_o$, $\mathfrak{M}_n$ is not empty. By the Axiom of Choice for Sequences, there exists a sequence $\langle V_n \rangle$ such that $V_n \in \mathfrak{M}_n$ for each $n \in I_o$. The set H is closed; therefore the set $V_0 = C(H)$ is open. Moreover, S is contained in the union of the countable collection of open sets consisting of V_0 and all the V_n. It is clear that no finite subcollection of this countable collection covers S. Therefore S is not countably compact. QED

We have now shown that in general metric spaces, compactness implies countable compactness, and countable compactness implies sequential compactness. Our goal is to prove that these three types of compactness are equivalent in metric spaces. Thus we would now like to prove that sequential compactness implies compactness. However, by virtue of Corollary 4.9, every open covering of a sequentially compact metric space has a countable open subcovering. Thus to prove that every sequentially compact metric space is compact, it is sufficient to prove that every sequentially compact metric space is countably compact. We do this by the Cantor Product Theorem, which we now establish.

Theorem 4.11 (Cantor Product Theorem). Let $\langle A_n \rangle$ be a sequence of subsets of a metric space S satisfying the following conditions:

(*a*) A_1 is sequentially compact.
(*b*) A_n is closed for each $n \in I_o$.
(*c*) $A_n \neq \phi$ for $n \in I_o$.
(*d*) $A_n \supset A_{n+1}$ for each $n \in I_o$.

Then $\bigcap_{n \in I_o} A_n$ is nonempty, closed, and sequentially compact.

Proof. The proof that the intersection is both closed and sequentially compact is left as an exercise. To prove that the intersection is nonempty, we note that it follows from (*c*) and the Axiom of Choice for Sequences that there exists a sequence of points $\langle p_n \rangle$ such that $p_n \in A_n$ for each $n \in I_o$. The sequence $\langle p_n \rangle$ is contained in the sequentially compact set A_1, and it must thus have a convergent subsequence $\langle p_{n_i} \rangle$. Define $p = \lim p_{n_i}$. We shall prove that $p \in \bigcap_{n \in I_o} A_n$. Suppose that this is not so. Then there exists an integer N such that p is not an element of the closed set A_N. This is a contradiction, since $p_{n_i} \in A_N$ for $i > N$. QED

Theorem 4.12. Let S be a metric space. Then if S has any one of the following three properties, it has all of them:

(*a*) Compact.
(*b*) Sequentially compact.
(*c*) Countably compact.

Furthermore, there exist metric spaces which are both closed and bounded, but which have none of these three properties.

Proof. From the previous discussion, it is sufficient to prove that if S is sequentially compact, then S is countably compact. Let $\{G_n\}$ be any countable open covering of the sequentially compact metric space S. Suppose that no finite subcollection of $\{G_n\}$ covers S. Define the following sets:

$$\begin{aligned} A_1 &= S - G_1 \\ A_2 &= S - (G_1 \cup G_2) \\ &\vdots \\ A_n &= S - (G_1 \cup G_2 \cup \ldots \cup G_n) \\ &\vdots \end{aligned}$$

By DeMorgan's Theorem, it follows that each A_i is a closed set, and hence each A_i is sequentially compact by Problem 32. From the fact that no finite subcollection of $\{G_n\}$ covers S, we see that each A_i is nonempty. It is clear that $A_n \supset A_{n+1}$ for every n. Accordingly, all the hypotheses of the Cantor Product Theorem are satisfied. Therefore, $\bigcap_{n \in I_o} A_n \neq \phi$. Let $p \in \bigcap_{n \in I_o} A_n$. This means that for every positive integer n,

$$p \in A_n = S - (G_1 \cup G_2 \cup \ldots \cup G_n)$$

so that

$$p \in S \qquad \text{and} \qquad p \notin G_n$$

This contradicts the fact that $S \subset \bigcup_{n \in I_0} G_n$. QED

PROBLEMS

35. Prove Theorem 4.5.
36. In the proof of Theorem 4.6, show that $p \in S(p_n, 1/j) \subset U$.
37. Prove the second half of Theorem 4.6.
38. Prove Corollary 4.9.
39. Fill in the missing details in the proof of the Cantor Product Theorem.
40. Fill in the missing details in the proof of Theorem 4.12.
41. Let A and B be sequentially compact subsets of a metric space S. Prove that $A \times B$ is sequentially compact. Deduce similar theorems for the other forms of compactness.

5. *Continuity, uniform continuity, and connectedness*

In this section, we define continuous mappings and uniformly continuous mappings from one metric space into another. We do this by generalizing the (ϵ, δ) definition which was discussed in R_1, and then we obtain several equivalent formulations of the definition of continuity similar to those which were derived in Chapter V. The definition of a connected subset of a metric space is then obtained in terms of continuous mappings, and an equivalent condition for connectedness is given. It is found that connectedness and all three forms of compactness are preserved under all continuous mappings. Applications of these results are given in the discussion of the diameter of a set and the distance between two sets. We also show that on compact subsets of a metric space, the properties of continuity and uniform continuity are equivalent.

Definition 5.1. Let S and T be metric sets (with metrics ρ_S and ρ_T, respectively), A a subset of S, and $f: A \to T$ a mapping.

(*a*) f is *continuous at the point* $p \in A$ iff given any $\epsilon > 0$, there exists a $\delta = \delta(p, \epsilon) > 0$ such that if $x \in A$ and $\rho_S(x, p) < \delta$, then $\rho_T(f(x), f(p)) < \epsilon$.

(*b*) f is *continuous on* A iff f is continuous at p for every $p \in A$.

(*c*) f is *uniformly continuous on* A iff given any $\epsilon > 0$, there exists a $\delta = \delta(\epsilon) > 0$ such that if $x_1, x_2 \in A$ and $\rho_S(x_1, x_2) < \delta$, then

$$\rho_T[f(x_1), f(x_2)] < \epsilon$$

In part (*a*) of Definition 5.1, the notation $\delta = \delta(p, \epsilon)$ means that δ is a function of both p and ϵ. A similar remark holds for the notation in part (*c*). It should be noted that Definition 5.1 holds equally well for metric spaces, since every metric space is a metric set.

We note by Example 3.1 of Chapter V that it is not true in general that continuity on a set implies uniform continuity on the same set. This is true, however, when the given set is compact, as we shall show in Theorem 5.27.

Let U be a subset of a metric space S. By Definition 1.6, the metric for U is a mapping $\rho: U \times U \to R_1$ satisfying certain conditions, in particular, the triangle inequality. We shall prove that the mapping ρ of the metric set $U \times U$ (see Theorem 1.10) into R_1 is uniformly continuous on $U \times U$.

Theorem 5.2. Let U be a subset of a metric space S with metric ρ. Let $A = U \times U$, and define the mapping $f: A \to R_1$ by $f(p) = \rho(x, y)$ for each point $p = (x, y)$ of A. Then f is uniformly continuous on A.

Proof. By Theorem 1.10, we see that the set A is a metric set with metric ρ_A defined by

$$(a) \qquad \rho_A(p_1, p_2) = \{[\rho(x_1, y_1)]^2 + [\rho(x_2, y_2)]^2\}^{1/2}$$

where $p_1 = (x_1, x_2)$ and $p_2 = (y_1, y_2)$ are two arbitrary points of A. It is evident from (*a*) that the following two properties hold:

$$(b) \quad \rho(x_1, y_1) \leq \rho_A(p_1, p_2) \qquad \text{and} \qquad \rho(x_2, y_2) \leq \rho_A(p_1, p_2)$$

To prove that f is uniformly continuous on A, let $\epsilon > 0$ be given. Define $\delta = \epsilon/2$. We shall prove that if p_1 and p_2 are any two points of A with $\rho_A(p_1, p_2) < \delta$, then $|f(p_1) - f(p_2)| < \epsilon$. This is equivalent to proving the following two assertions:

$$(c) \qquad f(p_1) - f(p_2) < \epsilon$$

$$(d) \qquad f(p_2) - f(p_1) < \epsilon$$

We prove (*c*) and leave the proof of (*d*) as an exercise.

To prove (*c*), let $p_1 = (x_1, x_2)$ and $p_2 = (y_1, y_2)$. Suppose that $\rho_A(p_1, p_2) < \delta = \epsilon/2$. By (*b*), this implies that

$$(e) \qquad \rho(x_1, y_1) < \frac{\epsilon}{2} \quad \text{and} \quad \rho(x_2, y_2) < \frac{\epsilon}{2}$$

From the definition of the mapping f, we have at once

$$\begin{aligned} f(p_1) - f(p_2) &= \rho(x_1, x_2) - \rho(y_1, y_2) \\ &\leq \rho(x_1, y_1) + \rho(y_1, y_2) + \rho(y_2, x_2) - \rho(y_1, y_2) \\ &= \rho(x_1, y_1) + \rho(x_2, y_2) \\ &< \frac{\epsilon}{2} + \frac{\epsilon}{2} \\ &= \epsilon \end{aligned}$$

QED

Corollary 5.3. Let A be a subset of a metric space S, and $a \in A$. Define a mapping $f\colon A \to R_1$ by $f(x) = \rho(x, a)$ for each $x \in A$. Then f is uniformly continuous on A.

Corollary 5.4. Let A and B be subsets of a metric space S, and $f\colon A \times B \to R_1$ defined by $f(p) = \rho(x, y)$ for each point $p = (x, y) \in A \times B$. Then f is uniformly continuous on $A \times B$.

Recall that in R_1 we defined continuity at a point in terms of sequences, and then obtained the equivalent (ϵ, δ) formulation as a theorem. In this section, we have reversed the procedure, taking the (ϵ, δ) formulation as our definition. We now prove the equivalent sequential formulation as a theorem.

Theorem 5.5. Let S and T be metric spaces, A a subset of S, $p \in A$, and $f\colon A \to T$ a mapping. Then f is continuous at p iff given any sequence of points $\langle p_n \rangle$ of A converging to p, the sequence $\langle f(p_n) \rangle$ converges to $f(p)$.

Proof. Half of the theorem follows from Problem 27. It remains to be proven that if the given condition holds, then f is continuous at p. Using the contrapositive technique, assume that f is not continuous at p. This means that there exists an $\epsilon_0 > 0$ such that, given any real number $\delta > 0$, we can find a point $x \in A$ with the property that $\rho_S(x, p) < \delta$ but $\rho_T[f(x), f(p)] \geq \epsilon_0$. For each positive integer n, define $\delta_n = 1/n$. We may thus find a point $x_n \in A$ such that $\rho_S(x_n, p) < 1/n$ but $\rho_T[f(x_n), f(p)] \geq \epsilon_0$. The student can easily show that the following conditions hold:

(*a*) The sequence $\langle x_n \rangle$ converges to p.

(*b*) The sequence $\langle f(x_n)\rangle$ does not converge to $f(p)$.
Therefore, f is not continuous at the point p. QED

Theorem 5.6. Let S and T be metric spaces, A a subset of S, $f\colon A \to T$ a continuous mapping, and $B = f(A)$. Then if A has any one of the following properties, B has all of them. In other words, each of these properties is preserved by all continuous mappings.

(*a*) A is compact.
(*b*) A is sequentially compact.
(*c*) A is countably compact.

The proof is left as an exercise, using the proof of Theorem 1.26 of Chapter V as a model.

Theorem 5.7. (*a*) Let A be a compact nonempty subset of a metric space S. Then there exist points $a, b \in A$ such that

$$\rho(x, y) \leq \rho(a, b)$$

for every pair of points $x, y \in A$.

(*b*) Let A and B be disjoint, nonempty, compact subsets of a metric space S. Then there exist points $a \in A$ and $b \in B$ such that

$$\rho(x, y) \geq \rho(a, b) \neq 0$$

for every $x \in A$, $y \in B$.

Proof. By Problem 41, the set $A \times B$ is compact. Let f be the mapping defined in Corollary 5.4. Then by Theorem 5.6, the subset $H = f(A \times B)$ of R_1 is compact. Consequently, H is both closed and bounded, and H contains its infimum m and its supremum M.

We now prove (*b*). We know that there exists a point $p = (a, b) \in A \times B$ such that $f(p) = m$, that is, $\rho(a, b) = m$. Since $a \in A$ and $b \in B$, $\rho(a, b) \neq 0$. By the definition of f, $\rho(x, y) \geq \rho(a, b)$ for every pair of points $x \in A$, $y \in B$. This proves (*b*).

The proof of (*a*) is left as an exercise, with the suggestion that the student take $B = A$. QED

In Theorem 5.7(*a*), we have seen that for any nonempty compact subset A of a metric space S, there exist two points a and b of A whose distance apart is a maximum. In the specific case where A is a circle, two points of A whose distance apart is a maximum are said to be endpoints of a diameter of A, and the diameter of A is the real number defined to be the distance between these two points. We now generalize this definition.

Definition 5.8. Let A be a bounded subset of a metric space S. The *diameter of* A, written $\delta(A)$, is the real number defined by the equations

(*a*) $\delta(A) = 0$ if $A = \phi$.
(*b*) $\delta(A) = \sup \{\rho(x, y): x, y \in A\}$ if $A \neq \phi$.

Theorem 5.9. Let A and B be bounded subsets of a metric space S with metric ρ. Then

(*a*) $\delta(A) = 0$ iff A contains at most one point.
(*b*) If $A \subset B$, then $\delta(A) \leq \delta(B)$.
(*c*) $\delta(\bar{A}) = \delta(A)$.
(*d*) If $A \cap B \neq \phi$, then $\delta(A \cup B) \leq \delta(A) + \delta(B)$.

Theorem 5.10. Let S be any metric space, ρ any metric for S, and A any compact subset of S. Then A is bounded, and there exist points $a, b \in A$ such that $\delta(A) = \rho(a, b)$.

The result of Theorem 5.7(*b*) leads naturally to the definition of the distance between two sets.

Definition 5.11. Let A and B be subsets of a metric space S with metric ρ. The *distance between* A *and* B, written $\rho(A, B)$, is the real number defined by the equations

(*a*) $\rho(A, B) = 0$ if either $A = \phi$ or $B = \phi$.
(*b*) $\rho(A, B) = \inf \{\rho(x, y): x \in A, y \in B\}$ if $A \neq \phi$ and $B \neq \phi$.

Theorem 5.12. Let A and B be bounded subsets of a metric space S with metric ρ. Then

(*a*) If $x \in A$, $y \in B$, then $0 \leq \rho(A, B) \leq \rho(x, y) \leq \delta(A \cup B)$.
(*b*) $\delta(A \cup B) \leq \delta(A) + \rho(A, B) + \delta(B)$.
(*c*) If $x \in S$, then $\rho(x, A) = 0$ iff $x \in \bar{A}$. This result holds even if A is unbounded.
(*d*) If the sets A and B are nonempty, compact, and disjoint, then $\rho(A, B) > 0$. Moreover, there exist points $a \in A$ and $b \in B$ such that $\rho(a, b) = \rho(A, B)$.

Example 5.13. Define the metric space S to consist of all points of R_1 with the exception of the points 2 and 3, and with the metric

$\rho(x, y) = |x - y|$, for $x, y \in S$. Define the subsets A and B of S as follows:

$$A = \left\{x : x \in S, x = 2 - \frac{1}{n}, n \in I_0\right\}$$

$$B = \left\{x : x \in S, x = 3 + \frac{1}{n}, n \in I_0\right\}$$

The student should show that A and B are closed and bounded subsets of S. Moreover, $\rho(A, B) = 1$. However, there do not exist points $a \in A$, $b \in B$ such that $\rho(a, b) = 1$. It should be noted that neither of the sets A, B is compact.

Example 5.14. Let the space be $R_2 = R_1 \times R_1$. Define A as the subset of R_2 consisting of all points on the X-axis, and B the subset of R_2 consisting of all points on the graph of the hyperbola $y = 1/x$. Clearly, A and B are disjoint, nonempty, closed sets, but $\rho(A, B) = 0$. Neither the set A nor the set B is compact.

Returning to our goal of obtaining several equivalent formulations of continuity in metric spaces, we next generalize the notion of a suburb of a point.

Definition 5.15. A subset K of a metric space S is a *suburb* of a point p of S iff $K = \bar{K}$ and $p \notin \bar{K}$.

Theorem 5.16. Let S and T be metric spaces, $A \subset S$, $p \in A$, and $f: A \to T$ a mapping. Then f is continuous at p iff for every suburb K of $f(p)$, $\overline{f^{-1}(K)}$ is a suburb of p.

The proof is modeled along the lines of that given for Theorem 2.7 of Chapter V, and is left as an exercise.

Definition 5.17. Let A and K be subsets of a metric space S. Then,

(*a*) K is said to be *closed in* A iff $K = \bar{K} \cap A$.

(*b*) K is said to be *open in* A iff $A - K$ is closed in A.

The proof of the next theorem is modeled along the lines of that given for Theorem 2.10 of Chapter V, and is left as an exercise.

Theorem 5.18. Let A and B be subsets of a metric space S, and $f: A \to B$ a mapping. Then, f is continuous iff either of the following conditions holds:

(*a*) If K is any subset of B which is closed in B, then $f^{-1}(K)$ is closed in A.

(*b*) If K is any subset of B which is open in B, then $f^{-1}(K)$ is open in A.

As our first application of Theorem 5.18, we give an alternate proof of Theorem 5.6.

Alternate proof of Theorem 5.6. We shall prove that if A is compact, then B is compact. Let $\{G_\alpha\}$ be any collection of open subsets of T which covers B. For each α, define $K_\alpha = G_\alpha \cap B$. By Problem 71, each of the sets K_α is open in B. For each α, define $H_\alpha = f^{-1}(K_\alpha)$. By Theorem 5.18, each of the sets H_α is open in A. Thus, for each α, there exists an open set U_α in S such that $H_\alpha = U_\alpha \cap A$. The collection $\{U_\alpha\}$ is an open covering of A. Since A is compact, this collection contains a finite open subcovering of A, which we denote by $\{U_{\alpha_1}, U_{\alpha_2}, \ldots, U_{\alpha_n}\}$. Therefore, $\{H_{\alpha_1}, H_{\alpha_2}, \ldots, H_{\alpha_n}\}$ is a finite covering of A. It follows from this that $\{K_{\alpha_1}, K_{\alpha_2}, \ldots, K_{\alpha_n}\}$ is a finite covering of B, from which we see at once that $\{G_{\alpha_1}, G_{\alpha_2}, \ldots, G_{\alpha_n}\}$ is a finite open covering of B. Therefore B is compact. QED

Our next application of Theorem 5.18 is in the proof of the composition theorem for continuous mappings.

Theorem 5.19. Let S, T, and U be metric spaces, A a subset of S, B a subset of T, $f: A \to B$, and $g: B \to U$ continuous mappings. Then the composition mapping $gf: A \to U$ defined by $gf(x) = g[f(x)]$ for each $x \in A$ is continuous.

Proof. We shall make use of Theorem 5.18(*a*). Let K be any subset of U which is closed in U. Since g is a continuous mapping, the subset $g^{-1}(K)$ of B is closed in B. Consequently, since f is continuous, the subset $f^{-1}[g^{-1}(K)]$ is closed in A. However, $f^{-1}[g^{-1}(K)] = (gf)^{-1}(K)$. Therefore, by Theorem 5.18(*a*), the mapping gf is continuous. QED

Our definition of connected sets in R_1 was based on the concept of "betweenness," which depends upon the ordering of the real numbers. Since we have not introduced any notion of order in general metric spaces, we prefer to define connected sets without using this concept. We take as our definition a restatement of Theorem 1.34 of Chapter V.

Definition 5.20. A subset H of a metric space S is *connected* iff there do not exist two distinct points a and b of R_1 and a continuous mapping f of H into R_1 such that $f(H) = \{a\} \cup \{b\}$.

Theorem 5.21. Let S and T be metric spaces, and A a connected subset of S. Let $f: A \to T$ be a continuous mapping. Then $f(A)$ is a connected subset of T.

The proof is left as an exercise.

Theorem 5.22. Let A be a connected subset of a metric space S, and suppose that A contains two distinct points. Then A is uncountable.

Proof. Let a and b be distinct points of A. Define a mapping $f: A \to R_1$ by $f(x) = \rho(x, a)$ for each $x \in A$. By Corollary 5.3, f is continuous. Therefore the set $f(A)$ is a connected subset of R_1. This set contains the point $f(a) = 0$ and the point $f(b) = \rho(b, a) = c > 0$. Consequently, since $f(A)$ is connected, it must contain the closed interval $[0, c]$. Therefore $f(A)$ is uncountable, so that A is uncountable. QED

We now give a characterization of connected sets that does not depend upon the notion of a mapping. This characterization is based on the following lemma, the proof of which is left as an exercise.

Lemma 5.23. Let a and b be two distinct points of R_1, and define $H = \{a\} \cup \{b\}$. Then each of the subsets $\{a\}$ and $\{b\}$ of H is both open and closed in H.

Theorem 5.24. A subset K of a metric space S is connected iff K cannot be written in the form $K = A \cup B$, where A and B are disjoint, nonempty sets, each of which is both open and closed in K.

The proof is a simple application of Lemma 5.23, Theorem 5.18, and the definition of a connected set.

Definition 5.25. A subset H of a set K is said to be a *proper subset* of K iff $H \neq \phi$ and $H \neq K$.

The next theorem is a corollary to Theorem 5.24.

Theorem 5.26. A subset K of a metric space S is connected iff no proper subset of K is both open and closed in K.

We close with a proof of the important fact that every continuous mapping defined on a compact set is uniformly continuous.

Theorem 5.27. Let S and T be metric spaces, A a compact subset of S, and $f: A \to T$ a continuous mapping. Then f is uniformly continuous on A.

Proof. Pick any $\epsilon > 0$. Given any point $p \in A$, we know that f is continuous at the point p. This means that there exists a $\delta_p > 0$ such that if x is any point of A satisfying $\rho_S(x, p) < \delta_p$, then $\rho_T[f(x), f(p)] < \epsilon/2$. For each point $p \in A$, define the spherical neighborhood $G_p = S(p, \frac{1}{2}\delta_p)$. Then $\{G_p\}$ is an open covering of the compact set A; hence there exists a finite open subcovering $\{G_{p_1}, G_{p_2}, \ldots, G_{p_k}\}$ of A. Corresponding to this finite subcovering is the finite set of positive numbers $\{\delta_{p_1}, \delta_{p_2}, \ldots, \delta_{p_k}\}$. Define $\delta = \frac{1}{2} \inf \{\delta_{p_1}, \delta_{p_2}, \ldots, \delta_{p_k}\}$. Let x and y be any two points of A such that $\rho_S(x, y) < \delta$. There exists an integer i_o such that $x \in G_{p_{i_o}}$, where $1 \leq i_o \leq k$. This means that

$$\text{(a)} \qquad \rho_S(x, p_{i_o}) < \tfrac{1}{2}\, \delta_{p_{i_o}}$$

Consequently

$$\text{(b)} \qquad \rho_T[f(x), f(p_{i_o})] < \frac{\epsilon}{2}$$

On the other hand,

$$\begin{aligned} \rho_S(y, p_{i_o}) &\leq \rho_S(y, x) + \rho_S(x, p_{i_o}) \\ &< \delta + \tfrac{1}{2}\delta_{p_{i_o}} \\ &\leq \tfrac{1}{2}\delta_{p_{i_o}} + \tfrac{1}{2}\delta_{p_{i_o}} \\ &= \delta_{p_{i_o}} \end{aligned}$$

This means that

$$\text{(c)} \qquad \rho_S(y, p_{i_o}) < \delta_{p_{i_o}}$$

so that

$$\text{(d)} \qquad \rho_T[f(y), f(p_{i_o})] < \frac{\epsilon}{2}$$

From (*b*) and (*d*), we have at once

$$\text{(e)} \qquad \rho_T[f(x), f(y)] < \epsilon$$

QED

PROBLEMS

42. Prove assertion (*b*) in Theorem 5.2.
43. Prove assertion (*d*) in Theorem 5.2.

44. Prove that a mapping which is uniformly continuous on a subset A of a metric space is continuous on A.
45. Prove Corollary 5.3.
46. Prove Corollary 5.4.
47. Let S and T be metric spaces, A a subset of S, $p \in A$, and $f: A \to T$ a mapping which is continuous at the point p. Let $\langle p_n \rangle$ be any sequence of points of A converging to the point p. Prove that the sequence $\langle f(p_n) \rangle$ converges to the point $f(p)$.
48. Let S and T be metric spaces, A a subset of S, and $f: A \to T$ a mapping which is uniformly continuous on A. Let $\langle p_n \rangle$ be any Cauchy sequence of points in A. Prove that $\langle f(p_n) \rangle$ is a Cauchy sequence of points of T.
49. Prove (*a*) and (*b*) of Theorem 5.5.
50. Prove Theorem 5.6.
51. Prove Theorem 5.7(*a*).
52. Prove that in a metric space S, every closed subset of a compact set is compact.
53. Prove that a subset H of $R_2 = R_1 \times R_1$ is compact iff H is both closed and bounded. (*Hint:* By R_2 we denote either of the equivalent metric spaces of Theorem 2.25. The fact that H is bounded implies that there exists a square in R_2 which is centered at the origin with sides parallel to the coordinate axes and which has H in its interior. Let B be the union of this square and its interior. Show that there exists a closed interval K in R_1 such that $B = K \times K$. Then make use of Problems 41 and 52.)
54. Prove Theorem 5.9(*a*).
55. Prove Theorem 5.9(*b*).
56. Prove Theorem 5.9(*c*).
57. Prove Theorem 5.9(*d*).
58. Prove Theorem 5.10.
59. Prove Theorem 5.12(*a*).
60. Prove Theorem 5.12(*b*).
61. Prove Theorem 5.12(*c*).
62. Prove Theorem 5.12(*d*).
63. Fill in the missing details of Example 5.13.
64. Fill in the missing details of Example 5.14.
65. Let A and B be nonempty, disjoint, closed subsets of $R_2 = R_1 \times R_1$, and suppose that B is bounded. Prove that there exist points $a \in A$, $b \in B$, such that $0 < \rho(a, b) = \rho(A, B)$.
66. A sequence of points $\langle p_n \rangle$ of a metric space S converges to the point p of S iff no suburb of p contains a subsequence of $\langle p_n \rangle$.
67. Prove Theorem 5.16.
68. Let A be any subset of a metric space S, and $\{G_\alpha\}$ any collection of sets, each of which is closed in A. Then the intersection of all the G_α is

closed in A, and the union of any finite subcollection of the G_α is closed in A.

69. Let A be any subset of a metric space S, and $\{G_\alpha\}$ any collection of sets, each of which is open in A. Then the union of all the G_α is open in A, and the intersection of any finite subcollection of the G_α is open in A.

70. Prove Theorem 5.18.

71. Let A be a subset of a metric space S, and G a subset of A. Then G is open in A iff there exists an open set H in S such that $G = H \cap A$.

72. In the proof of Theorem 5.19, prove that $f^{-1}[g^{-1}(K)] = (gf)^{-1}(K)$.

73. Give an alternate proof of Theorem 5.19 using Theorem 5.18(*b*).

74. Give an alternate proof of Theorem 5.19 using the sequential definition of continuity.

75. Give an alternate proof of Theorem 5.19 using the (ϵ, δ) definition of continuity.

76. Prove Theorem 5.21. (*Hint:* See the proof of Theorem 1.35 of Chapter V.)

77. Prove that if $\{G_\alpha\}$ is a collection of connected subsets of a metric space S such that no two of the G_α are disjoint, then the union of the G_α is connected.

78. Prove Lemma 5.23.

79. Prove Theorem 5.24.

80. Prove Theorem 5.26.

81. Prove part (*e*) of Theorem 5.27.

82. Let A and B be subsets of a metric space S, each of which is closed in $A \cup B$. Suppose that each of the sets $A \cup B$ and $A \cap B$ is connected. Prove that A is connected and B is connected.

Index of Axioms and Key Theorems

Index of axioms

Axiom of Choice for Sequences, 32
Least Upper Bound Axiom, 26
Well-Ordering Axiom for I_o, 32

Index of key theorems

(In order of first occurrence)

DeMorgan's laws, 9, 67, 150
Archimedean property of R_1, 27
Every sequence of real numbers has a monotone subsequence, 34
A set of real numbers is unbounded if and only if it contains a monotone subsequence which is uniformly isolated, 37
A set H is infinite if and only if H contains a sequence of distinct points, 45
Catalogue of connected sets in R_1, 55
A subset H of R_1 is open if and only if H is the union of a countable collection of open intervals, 63
R_1 is not the union of two nonempty disjoint open sets, 65
Heine-Borel Theorem, 69, 71
Every bounded monotone sequence of real numbers converges, 76
Bolzano-Weierstrass Theorem for Sequences, 80
A sequence of real numbers is convergent if and only if it is a Cauchy sequence, 89

Index

(See also Index of Axioms and Key Theorems)

A CATALOG OF SELECTED

DOVER BOOKS

IN SCIENCE AND MATHEMATICS

Mathematics

FUNCTIONAL ANALYSIS (Second Corrected Edition), George Bachman and Lawrence Narici. Excellent treatment of subject geared toward students with background in linear algebra, advanced calculus, physics and engineering. Text covers introduction to inner-product spaces, normed, metric spaces, and topological spaces; complete orthonormal sets, the Hahn-Banach Theorem and its consequences, and many other related subjects. 1966 ed. 544pp. 6⅛ x 9¼. 0-486-40251-7

DIFFERENTIAL MANIFOLDS, Antoni A. Kosinski. Introductory text for advanced undergraduates and graduate students presents systematic study of the topological structure of smooth manifolds, starting with elements of theory and concluding with method of surgery. 1993 edition. 288pp. 5⅜ x 8½. 0-486-46244-7

VECTOR AND TENSOR ANALYSIS WITH APPLICATIONS, A. I. Borisenko and I. E. Tarapov. Concise introduction. Worked-out problems, solutions, exercises. 257pp. 5⅝ x 8¼. 0-486-63833-2

AN INTRODUCTION TO ORDINARY DIFFERENTIAL EQUATIONS, Earl A. Coddington. A thorough and systematic first course in elementary differential equations for undergraduates in mathematics and science, with many exercises and problems (with answers). Index. 304pp. 5⅜ x 8½. 0-486-65942-9

FOURIER SERIES AND ORTHOGONAL FUNCTIONS, Harry F. Davis. An incisive text combining theory and practical example to introduce Fourier series, orthogonal functions and applications of the Fourier method to boundary-value problems. 570 exercises. Answers and notes. 416pp. 5⅜ x 8½. 0-486-65973-9

COMPUTABILITY AND UNSOLVABILITY, Martin Davis. Classic graduate-level introduction to theory of computability, usually referred to as theory of recurrent functions. New preface and appendix. 288pp. 5⅜ x 8½. 0-486-61471-9

AN INTRODUCTION TO MATHEMATICAL ANALYSIS, Robert A. Rankin. Dealing chiefly with functions of a single real variable, this text by a distinguished educator introduces limits, continuity, differentiability, integration, convergence of infinite series, double series, and infinite products. 1963 edition. 624pp. 5⅜ x 8½. 0-486-46251-X

METHODS OF NUMERICAL INTEGRATION (SECOND EDITION), Philip J. Davis and Philip Rabinowitz. Requiring only a background in calculus, this text covers approximate integration over finite and infinite intervals, error analysis, approximate integration in two or more dimensions, and automatic integration. 1984 edition. 624pp. 5⅜ x 8½. 0-486-45339-1

INTRODUCTION TO LINEAR ALGEBRA AND DIFFERENTIAL EQUATIONS, John W. Dettman. Excellent text covers complex numbers, determinants, orthonormal bases, Laplace transforms, much more. Exercises with solutions. Undergraduate level. 416pp. 5⅜ x 8½. 0-486-65191-6

RIEMANN'S ZETA FUNCTION, H. M. Edwards. Superb, high-level study of landmark 1859 publication entitled "On the Number of Primes Less Than a Given Magnitude" traces developments in mathematical theory that it inspired. xiv+315pp. 5⅜ x 8½. 0-486-41740-9

Math–Decision Theory, Statistics, Probability

INTRODUCTION TO PROBABILITY, John E. Freund. Featured topics include permutations and factorials, probabilities and odds, frequency interpretation, mathematical expectation, decision-making, postulates of probability, rule of elimination, much more. Exercises with some solutions. Summary. 1973 edition. 247pp. $5\frac{3}{8}$ x $8\frac{1}{2}$. 0-486-67549-1

STATISTICAL AND INDUCTIVE PROBABILITIES, Hugues Leblanc. This treatment addresses a decades-old dispute among probability theorists, asserting that both statistical and inductive probabilities may be treated as sentence-theoretic measurements, and that the latter qualify as estimates of the former. 1962 edition. 160pp. $5\frac{3}{8}$ x $8\frac{1}{2}$. 0-486-44980-7

APPLIED MULTIVARIATE ANALYSIS: Using Bayesian and Frequentist Methods of Inference, Second Edition, S. James Press. This two-part treatment deals with foundations as well as models and applications. Topics include continuous multivariate distributions; regression and analysis of variance; factor analysis and latent structure analysis; and structuring multivariate populations. 1982 edition. 692pp. $5\frac{3}{8}$ x $8\frac{1}{2}$. 0-486-44236-5

LINEAR PROGRAMMING AND ECONOMIC ANALYSIS, Robert Dorfman, Paul A. Samuelson and Robert M. Solow. First comprehensive treatment of linear programming in standard economic analysis. Game theory, modern welfare economics, Leontief input-output, more. 525pp. $5\frac{3}{8}$ x $8\frac{1}{2}$. 0-486-65491-5

PROBABILITY: AN INTRODUCTION, Samuel Goldberg. Excellent basic text covers set theory, probability theory for finite sample spaces, binomial theorem, much more. 360 problems. Bibliographies. 322pp. $5\frac{3}{8}$ x $8\frac{1}{2}$. 0-486-65252-1

GAMES AND DECISIONS: INTRODUCTION AND CRITICAL SURVEY, R. Duncan Luce and Howard Raiffa. Superb nontechnical introduction to game theory, primarily applied to social sciences. Utility theory, zero-sum games, n-person games, decision-making, much more. Bibliography. 509pp. $5\frac{3}{8}$ x $8\frac{1}{2}$. 0-486-65943-7

INTRODUCTION TO THE THEORY OF GAMES, J. C. C. McKinsey. This comprehensive overview of the mathematical theory of games illustrates applications to situations involving conflicts of interest, including economic, social, political, and military contexts. Appropriate for advanced undergraduate and graduate courses; advanced calculus a prerequisite. 1952 ed. x+372pp. $5\frac{3}{8}$ x $8\frac{1}{2}$. 0-486-42811-7

FIFTY CHALLENGING PROBLEMS IN PROBABILITY WITH SOLUTIONS, Frederick Mosteller. Remarkable puzzlers, graded in difficulty, illustrate elementary and advanced aspects of probability. Detailed solutions. 88pp. $5\frac{3}{8}$ x $8\frac{1}{2}$. 0-486-65355-2

PROBABILITY THEORY: A CONCISE COURSE, Y. A. Rozanov. Highly readable, self-contained introduction covers combination of events, dependent events, Bernoulli trials, etc. 148pp. $5\frac{5}{8}$ x $8\frac{1}{4}$. 0-486-63544-9

THE STATISTICAL ANALYSIS OF EXPERIMENTAL DATA, John Mandel. First half of book presents fundamental mathematical definitions, concepts and facts while remaining half deals with statistics primarily as an interpretive tool. Well-written text, numerous worked examples with step-by-step presentation. Includes 116 tables. 448pp. $5\frac{3}{8}$ x $8\frac{1}{2}$. 0-486-64666-1

Physics

OPTICAL RESONANCE AND TWO-LEVEL ATOMS, L. Allen and J. H. Eberly. Clear, comprehensive introduction to basic principles behind all quantum optical resonance phenomena. 53 illustrations. Preface. Index. 256pp. 5⅜ x 8½. 0-486-65533-4

QUANTUM THEORY, David Bohm. This advanced undergraduate-level text presents the quantum theory in terms of qualitative and imaginative concepts, followed by specific applications worked out in mathematical detail. Preface. Index. 655pp. 5⅜ x 8½. 0-486-65969-0

ATOMIC PHYSICS (8th EDITION), Max Born. Nobel laureate's lucid treatment of kinetic theory of gases, elementary particles, nuclear atom, wave-corpuscles, atomic structure and spectral lines, much more. Over 40 appendices, bibliography. 495pp. 5⅜ x 8½. 0-486-65984-4

A SOPHISTICATE'S PRIMER OF RELATIVITY, P. W. Bridgman. Geared toward readers already acquainted with special relativity, this book transcends the view of theory as a working tool to answer natural questions: What is a frame of reference? What is a "law of nature"? What is the role of the "observer"? Extensive treatment, written in terms accessible to those without a scientific background. 1983 ed. xlviii+172pp. 5⅜ x 8½. 0-486-42549-5

AN INTRODUCTION TO HAMILTONIAN OPTICS, H. A. Buchdahl. Detailed account of the Hamiltonian treatment of aberration theory in geometrical optics. Many classes of optical systems defined in terms of the symmetries they possess. Problems with detailed solutions. 1970 edition. xv + 360pp. 5⅜ x 8½. 0-486-67597-1

PRIMER OF QUANTUM MECHANICS, Marvin Chester. Introductory text examines the classical quantum bead on a track: its state and representations; operator eigenvalues; harmonic oscillator and bound bead in a symmetric force field; and bead in a spherical shell. Other topics include spin, matrices, and the structure of quantum mechanics; the simplest atom; indistinguishable particles; and stationary-state perturbation theory. 1992 ed. xiv+314pp. 6⅛ x 9¼. 0-486-42878-8

LECTURES ON QUANTUM MECHANICS, Paul A. M. Dirac. Four concise, brilliant lectures on mathematical methods in quantum mechanics from Nobel Prize-winning quantum pioneer build on idea of visualizing quantum theory through the use of classical mechanics. 96pp. 5⅜ x 8½. 0-486-41713-1

THIRTY YEARS THAT SHOOK PHYSICS: THE STORY OF QUANTUM THEORY, George Gamow. Lucid, accessible introduction to influential theory of energy and matter. Careful explanations of Dirac's anti-particles, Bohr's model of the atom, much more. 12 plates. Numerous drawings. 240pp. 5⅜ x 8½. 0-486-24895-X

ELECTRONIC STRUCTURE AND THE PROPERTIES OF SOLIDS: THE PHYSICS OF THE CHEMICAL BOND, Walter A. Harrison. Innovative text offers basic understanding of the electronic structure of covalent and ionic solids, simple metals, transition metals and their compounds. Problems. 1980 edition. 582pp. 6⅛ x 9¼. 0-486-66021-4